T0358262

LARVAE OF ANOMURAN AND BRACHYURAN CRABS
OF NORTH CAROLINA

A GUIDE TO THE DESCRIBED LARVAL STAGES OF ANOMURAN
(FAMILIES PORCELLANIDAE, ALBUNEIDAE, AND HIPPIDAE)
AND BRACHYURAN CRABS OF NORTH CAROLINA, U.S.A.

Published in this series:

CRM 001 - *Stephan G. Bullard*   Larvae of anomuran and brachyuran crabs of North Carolina

In preparation (provisional titles):

CRM 002 - *Spyros Sfenthourakis et al.*   The biology of terrestrial isopods (V)
CRM 003 - *Tomislav Karanovic*   Subterranean Copepoda from arid Western Australia
CRM 004 - *Darren C. Yeo & Peter K. L. Ng*   The freshwater crabs of Indo China
CRM 005 - *Katsushi Sakai*   Callianassidae of the world

Author's address:

Wake Forest University, Department of Biology, P.O. Box 7325, Winston-Salem, NC 27109, U.S.A. Current address: University of Connecticut, Department of Marine Sciences, Marine Sciences Building, 1080 Shennecossett Rd., Groton, CT 06340, U.S.A.; e-mail: stephan.bullard@uconn.edu

Manuscript first received 6 December 2000; final version accepted 7 December 2001.

Cover: Emerita talpoida, zoea IV; see p. 68, fig. 9a.

LARVAE OF ANOMURAN AND BRACHYURAN CRABS OF NORTH CAROLINA

A GUIDE TO THE DESCRIBED LARVAL STAGES OF ANOMURAN (FAMILIES PORCELLANIDAE, ALBUNEIDAE, AND HIPPIDAE) AND BRACHYURAN CRABS OF NORTH CAROLINA, U.S.A.

BY

Stephan G. Bullard

CRUSTACEANA MONOGRAPHS, 1

BRILL

LEIDEN · BOSTON

ISBN 90 04 12841 7

CONTENTS

INTRODUCTION

This guide is a compilation of the available descriptions of the zoeal and megalopal stages of anomuran (families: Porcellanidae, Albuneidae, and Hippidae) and brachyuran crabs of North Carolina, U.S.A. Descriptions of the zoeae of 45 species of crabs and the megalopae of 34 species are included. Keys are provided for the identification of these stages.

Keys to larval crabs are uncommon (Paula, 1996) with Sandifer's (1972) guide currently being the most comprehensive reference available for larval crabs of the east coast of North America. Although the Sandifer (1972) guide is an excellent resource for larval crab taxonomy, larvae of numerous additional species have been described since its publication. The focus of the present study has been to assimilate the currently published descriptions of zoeal and megalopal stages of crabs found off of North Carolina, U.S.A. and provide a key to their identification. To generate the species pool for this study, geographical ranges for species described by Williams (1984) have been adopted. It should be noted, however, that because many species of crabs have their northernmost or southernmost distribution off North Carolina, the information contained in this guide is applicable to a significant portion of the east coast of North America.

A major aim of the present key has been to use macroscopic features (external features visible with a dissecting microscope without the need for dissection) to identify crab larvae to the species level. While in many cases this has been possible, some taxa, like the Majidae, have very similar macroscopic morphologies. Thus, information has been provided to enable investigators to use smaller scale features to identify some taxa.

2

SPECIES LIST

The following species are covered by this guide. Taxonomic affiliations are based upon Williams (1984).

Infraorder ANOMURA
  Family Porcellanidae

  Family Albuneidae
  Family Hippidae

Infraorder BRACHYURA
  Family Dromiidae
  Family Calappidae
  Family Leucosiidae
  Family Majidae
  Family Cancridae
  Family Polybiinae
  Family Portunidae

LARVAL CRAB ANATOMY

Figs. 1, 2 and 3 provide a working vocabulary of larval crab taxonomy. For further information investigators are referred to Williams (1984) for an excellent glossary of carcinological terms and to Ingle (1992) for a more thorough treatment of larval crab morphology. Because more than one term can apply to the same morphological feature, the following definitions explain how terms are used in the present key.

Definitions

Carapace. — The covering of the cephalothorax.

Carapace spines. — Porcellanid-like larvae possess rostral and posterior spines. Brachyuran-like larvae possess rostral, dorsal and lateral spines. Some or all of these spines, however, may be absent or reduced in some species.

Chromatophore. — Contractile pigment bearing cell (not black). Chromatophores can be various colors, often being white, red, or orange. In larval descriptions where the color of chromatophores is not listed, the colors have not been described.

Melanophore. — Black contractile pigment bearing cell.

Pereiopods. — The walking legs. In zoeae, the appendages that will become the walking legs in later developmental stages.

Swimming setae. — Long setae on the exopod of each maxilliped. The term swimming setae is equivalent to natatory setae. These setae often increase in number during larval development and are very useful for distinguishing among zoeal stages within a particular species.

Telson. — The last element of the abdomen. The telson of most brachyuran zoeae is a 2-pronged fork (which is often referred to as bifurcate in the literature). In many species inner spines lie between the two prongs of the fork in an area that is sometimes referred to as the median telson arch. These spines are labeled by indicating the number of spines present on each prong of the telson. Hence, 3 + 3 would indicate 3 pairs of inner spines with 3 spines present on each prong. For example, the zoea V stage of *Ovalipes ocellatus* (fig. 1b) has 5 inner spines on each prong, or 5 + 5 inner spines.

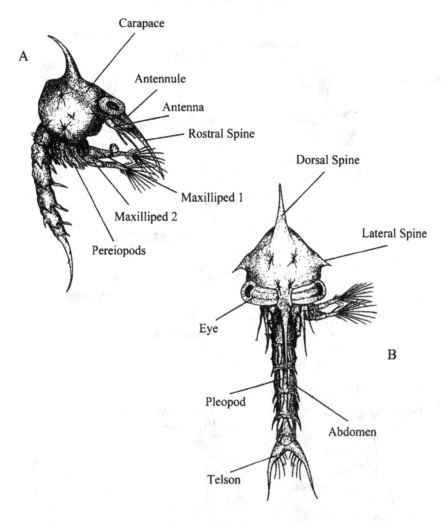

Fig. 1. Anatomy of a crab zoea. A, side view of *Ovalipes ocellatus* zoea V; B, ventral view of *Ovalipes ocellatus* zoea V. [Reprinted with permission from Costlow & Bookhout, 1966b.]

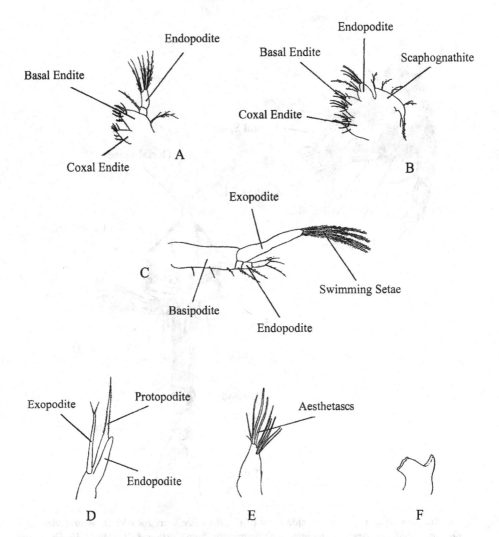

Fig. 2. Anatomy of a crab zoea. A, maxillule, *Ovalipes ocellatus* zoea I; B, maxilla, *O. ocellatus* zoea I; C, maxilliped 2, *O. ocellatus* zoea I; D, antenna, *O. ocellatus* zoea V; E, antennule, *O. ocellatus* zoea V; F, mandible, *O. ocellatus* zoea I. [Reprinted with permission from Costlow & Bookhout, 1966b.]

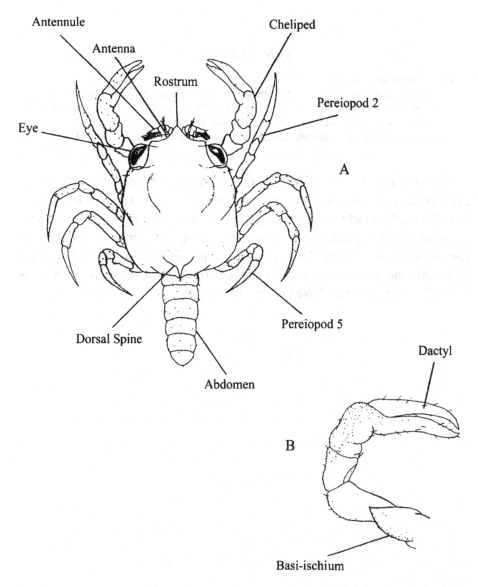

Fig. 3. Anatomy of a crab megalopa. A, dorsal view; B, cheliped of *Persephona mediterranea*. [Reprinted with permission from Negreiros-Fransozo et al., 1989.]

KEY TO ZOEAE

1. Overall appearance of zoea
   - Carapace triangular. Very long and straight rostral and posterior spines. No dorsal spine. Telson blade-like. E.g., *Porcellana sigsbeiana* (fig. 7) .... **Porcellanid-like key**
   - Carapace more or less rounded. Rostral, dorsal and postero-lateral spines may or may not be present. Telson usually bifurcate. E.g., *Hexapanopeus angustifrons* (figs. 32a, b) ....
   .... **Brachyuran-like key**

Note: The morphology of *Emerita talpoida* zoeae is neither completely porcellanid-like nor brachyuran-like. For purposes of this key, *E. talpoida* has been included in the brachyuran-like subgroup. Additionally, *Hypoconcha arcuata* zoeae are morphologically very similar to zoeae of some anomuran families not covered by the present key. Therefore, it is not possible to distinguish *H. arcuata* from other anomuran larvae based upon the information presented here. Descriptions of *H. arcuata* larvae have been included for the sake of completeness.

PORCELLANID-LIKE ZOEA KEY

1. Hooks on abdominal somite 5
   - Large posteriorly curving hooks on abdominal somite 5...............*Lepidopa* sp. (p. 19, fig. 8)
   - Large hooks absent from abdominal somite 5.. 2
2. Rostral spine spinules
   - Two ventral rows of spinules on rostral spine...............*Euceramus praelongus* (p. 16, fig. 4)
   - More than 2 rows of spinules on rostral spine, not just ventral.. 3
3. Hooks on basipodites of maxillipeds
   - Hooked spines on basipodites of maxillipeds..................*Porcellana sigsbeiana* (p. 18, fig. 7)
   - No hooked spines on basipodites 4
4. Spinules on posterior spines
   - Numerous spinules on posterior spines .... *Polyonyx gibbesi* (p. 18, fig. 6)
   - Very few (1 to 3) spinules on posterior spines .... *Megalobrachium soriatum* (p. 17, fig. 5)

BRACHYURAN-LIKE ZOEA KEY

1. Carapace spines
   - Carapace without dorsal, lateral, or rostral spine. Overall appearance of zoea shrimp-like .....
   .... *Hypoconcha arcuata* (p. 20, fig. 10)
   - Carapace without dorsal or lateral spines. Rostral spine very reduced. Carapace strongly rounded .... *Pinnotheres chamae* or *P. ostreum* (p. 44, figs. 37a and 38a, respectively) (see description of *P. chamae* to better differentiate between these species; see also *Emerita talpoida* zoea I, p. 19, fig. 9a)
   - Carapace with rostral and postero-lateral spines, but without dorsal spine ....
   .... *Emerita talpoida* (p. 19, figs. 9a, b)
   - Carapace with dorsal spine, but without lateral or rostral spines...
   .... *Anasimus latus* (p. 23, fig. 13)
   - Carapace with dorsal and rostral spines, but without lateral spines.. 2
   - Carapace with dorsal, rostral and lateral spines.. 3
2. Rostral spine length relative to length of antennae
   - Rostral spine longer than antennae... *Uca* spp. (p. 51, figs. 48a, b)
   - Rostral spine as long as antennae...
   .... *Sesarma cinereum* or *S. reticulatum* (pp. 48, 50, figs. 45a, b and 46a, respectively) (see description of *S. cinereum* to differentiate between these species)
   - Rostral spine shorter than antennae .... Majidae (pp. 23-27, figs. 13-18)
   (see description of *Anasimus latus* and Appendix I to differentiate among these species)
3. Lateral spine length
   - Lateral spines long and extending well beyond ventral margin of carapace........................... 4
   - Lateral spines short 8
4. Shape of abdominal somite 5
   - Abdominal somite 5 not wing shaped, but similar in appearance to other somites 5
   - Abdominal somite 5 wing shaped 7
5. Shape of the telson and size of antennae
   - Telson fluke-like. Antennae greatly reduced.... *Persephona mediterranea* (p. 22, figs. 12a, b)
   - Telson not fluke-like. Antennae not reduced.. 6
6. Relative size of carapace spines
   - Dorsal spine longest...*Pinnotheres maculatus* (p. 45, figs. 39a, b)
   - Rostral spine as long as, or longer than, dorsal spine .....*Dissodactylus* sp. (pp. 42, 43, figs. 35a, 36) (see description of *Dissodactylus crinitichelis* for more information about the genus)
7. Telson armature and prong width
   - Telson with median tooth on posterior margin..
   .... *Pinnixa chaetopterana* or *P. cristata* (pp. 46, 47, figs. 40, 41 and 44, respectively) (see description of *P. chaetopterana* to differentiate between these species)
   - Telson without median tooth, distance between the tips of telson prongs ≥ telson width ....
   ....*Pinnixa cylindrica* (p. 47, figs. 42, 44)
   - Telson without median tooth, distance between the tips of telson prongs < telson width ....
   .... *Pinnixa sayana* (p. 48, figs. 43, 44)
8. Frontal carapace protuberance
   - Zoea caltrop-shaped with distinct rounded protuberance on the central frontal region of the carapace...*Ocypode quadrata* (p. 51, figs. 47a, b)
   - No rounded protuberance on central frontal region of the carapace................................... 9

9. Presence of secondary rostral spines
 - Secondary rostral spines present, telson with two pairs of lateral spines, one small, one minute .... *Pseudomedaeus agassizii* (p. 34, figs. 26a, b)
 - Secondary rostral spines absent. Alternatively, secondary rostral spines present, telson with one pair of lateral spines... 10
10. Armature of abdominal somites
 - Small bifid postero-lateral spines on somites 3–5 .... *Arenaeus cribrarius* (p. 30, fig. 22)
 - Processes on somites 3–5 not bifid (rarely postero-lateral processes may appear bifid, but if so antennae and rostral spine are long and extend beyond the maxillipeds)........................ 11
11. Relative lengths of rostral spine, antennae and antennules
 - Antennae short, about half the length of rostral spine ....
   .... *Menippe mercenaria* (p. 41, figs. 34a, b)
 - Rostral spine shorter than antennae........................... *Pilumnus dasypodus* (p. 41, figs. 33a, b)
 - Rostral spine slightly longer than or equal in length to antennae, antennules about half the length of antennae (occasionally antennules may appear to be much less than half the length of antennae, but if so rostral spine relatively short and does not extend beyond the maxillipeds, and two rows of spines present on distal portion of antennae).................................... 12
 - Rostral spine slightly longer than or equal in length to antennae, antennules much less than half length of antennae 15
12. Armature of abdominal somites
 - Postero-lateral processes (knobs or spines) on all abdominal somites, but no lateral processes .... *Cancer irroratus* or *Cancer borealis* (pp. 27, 28, figs. 19a, b and 20a, b, respectively) (see description of *C. irroratus* to differentiate between these two species)
 - Lateral knob on abdominal somite 2. Postero-lateral spines on somites 3–5. Additional lateral hook on abdominal somite 5 and sometimes on somites 3–4..
   .... *Ovalipes ocellatus* (p. 29, figs. 21a, b)
 - A single lateral knob or hook on abdominal somites 2 and 3. Postero-lateral spines on abdominal somites 3–5 13
13. Spination of antennae
 - One row of spines on distal portion of antennae .... *Hepatus epheliticus* (p. 21, figs. 11a, b)
 - Two rows of spines on distal portion of antennae 14
14. Chromatophore pattern
 - Single chromatophore on dorsal spine........... *Portunus spinicarpus* (p. 33, figs. 25a, b, and c)
 - No chromatophore on dorsal spine ....
   Callinectes sapidus or *Callinectes similis* (pp. 31, 32, figs. 23a, b and c, and 24a, b and c, respectively) (see description of *C. sapidus* to differentiate between these two species)
15. Armature of abdominal somites
 - Very long postero-lateral spines on abdominal somite 5. No postero-lateral spines on other abdominal somites... *Rhithropanopeus harrisii* (p. 36, figs. 27a, b)
 - Lateral spines on abdominal somites 2 and 3. Postero-lateral spines on abdominal somites 3–5 (these may be reduced in early zoeal stages).. 16
16. Telson armature (1)
 - Telson with one small pair of lateral spines .... *Micropanope sculptipes* (p. 36, figs. 28a, b)
 - Telson without lateral spines.. 17
 - Telson with two pairs of lateral spines, one large, one small 19

KEY TO MEGALOPAE

Descriptions are not available, or poorly developed, for the megalopae of *Lepidopa* sp., *Stenocionops furcatus coelatus, Arenaeus cribrarius, Rhithropanopeus harrisii, Neopanope sayi, Menippe mercenaria, Dissodactylus mellitae, Pinnixa chaetopterana, Pinnixa cristata, Pinnixa cylindrica* and *Pinnixa sayana*. Additionally, *Euceramus praelongus, Megalobrachium soriatum, Polyonyx gibbesi, Porcellana sigsbeiana* and *Emerita talpoida* have megalopae that resemble the adult crabs and are not included in the following key.

1. Carapace spines
   - Carapace with prominent dorsal spine.. 2
   - Carapace without prominent dorsal spine 6
2. Anteriorly projecting carapace spines
   - Two anteriorly projecting carapace spines near the base of the eyes................................ 3
   - No anteriorly projecting carapace spines near the base of the eyes................................... 4
3. Hairs on the dactyl of pereiopod 5
   - Three stiff hairs on dactyl of pereiopod 5.................... *Pinnotheres maculatus* (p. 45, fig. 39b)
   - No stiff hairs on dactyl of pereiopod 5 .... *Anasimus latus* (p. 23, fig. 13)
4. Hairs on the dactyl of pereiopod 5
   - Three stiff hairs on dactyl of pereiopod 5.. 5
   - No stiff hairs on dactyl of pereiopod 5 .... *Persephona mediterranea* (p. 22, fig. 12b)
5. Setation of pleopods
   - Pleopods of somites 2–5 bear 16, 16, 16, 16 and 10 plumose setae, respectively ....
   .... *Cancer borealis* (p. 28, fig. 20b)
   - Pleopods of somites 2–5 bear 15, 15, 13, 11 and 9 plumose setae, respectively ....
   ....*Cancer irroratus* (p. 27, fig. 19b)
6. Spines on basi-ischium of chelipeds
   - Prominent spine on basi-ischium of each cheliped... 7
   - No spine on basi-ischium of chelipeds... 12
7. Posterior carapace spines
   - Pair of spines projecting from the ventral edge of carapace extending beyond pereiopod 5 ....
   8
   - No posterior carapace spines.. 9
8. Number of antennal segments
   - Antennae composed of 10 segments .... *Portunus spinicarpus* (p. 33, fig. 25c)
   - Antennae composed of 11 segments ....
   .... *Callinectes sapidus* or *Callinectes similis* (pp. 31, 32, figs. 23c and 24c,
   respectively) (see description of *C. sapidus* to differentiate between these two species)
9. Presence of rostral horns
   - Rostral horns present, rostrum depressed .... *Hexapanopeus angustifrons* (p. 40, fig. 32b)
   - Rostral horns absent. Alternatively, rostral horns present, rostrum not depressed 10

10. Hairs and morphology of dactyl of pereiopod 5
  – Three stiff hairs on dactyl of pereiopod 5. Dactyl of pereiopod 5 similar to other pereiopods
  11
  – Three stiff hairs on dactyl of pereiopod 5. Dactyl of pereiopod 5 flattened ....
  .... *Ovalipes ocellatus* (p. 29, fig. 21b)
  – No stiff hairs on dactyl of pereiopod 5 17
11. Number of setae between rostrum and lateral horns, and antennule morphology
  – Four long plumose setae between rostrum and each lateral horn ....
  .... *Micropanope sculptipes* (p. 36, fig. 28b)
  – One seta between rostrum and each lateral horn; 3 large aesthetascs on each side of the base
  of flagellum on the antennule ....*Panopeus herbstii* (p. 39, fig. 31b)
  – Two setae between rostrum and each lateral horn; no aesthetascs at the base of flagellum on
  the antennule .... *Pseudomedaeus agassizii* (p. 34, fig. 26b)
12. Rostrum
  – Rostrum visible from dorsal surface (possibly small) 13
  – Rostrum not visible from dorsal surface 18
13. Rostrum morphology
  – Rostrum long, pointed and directed anteriorly ....*Sesarma reticulatum* (p. 50, fig. 46b)
  – Rostrum long, directed downwards and strongly hook-like ....
  .... *Microphrys bicornutus* (p. 25, fig. 16)
  – Rostrum neither very long nor hook-like 14
14. Carapace tubercles (protuberances)
  – Carapace with tubercles 15
  – Carapace without tubercles 16
15. Arrangement of carapace tubercles
  – Pair of median cardiac tubercles ....*Libinia emarginata* (p. 24, fig. 15)
  – Single median cardiac tubercle and a mesogastric tubercle ....
  .... *Mithrax pleuracanthus* (p. 26, fig. 17)
  – Single median cardiac tubercle, but no mesogastric tubercle ......*Libinia dubia* (p. 24, fig. 14)
16. Carapace shape
  – Carapace sub-ovate in profile .... *Dissodactylus crinitichelis* (p. 42, fig. 35b)
  – Carapace not sub-ovate in profile .... *Hepatus epheliticus* (p. 21, fig. 11b)
17. Telson armature
  – Total of 7 terminal setae on telson ....*Eurypanopeus depressus* (p. 37, fig. 29b)
  – No terminal setae on telson ....*Pilumnus dasypodus* (p. 41, fig. 33b)
18. Placement of pereiopods
  – Pereiopod 4 carried dorsal to carapace ....*Ocypode quadrata* (p. 51, fig. 47b)
  – Pereiopod 5 carried dorsal to carapace. Carapace covered with hairs ....
  .... *Hypoconcha arcuata* (p. 20, fig. 10)
  – All pereiopods carried beneath carapace. Alternatively, pereiopod 5 carried dorsal to cara-
  pace, but carapace not covered with hairs 19

19. Hairs on the dactyl of pereiopod 5 and antennae composition
 - Three long hairs on dactyl of pereiopod 5, antennae composed of 9 segments.......................
  .... *Sesarma cinereum* (p. 48, fig. 45b)
 - No long hairs on dactyl of pereiopod 5. Alternatively, hairs on dactyl of pereiopod 5, anten-
   nae composed of more than 9 segments.. 20
20. Number of antennal segments
 - Antennae composed of 5 segments 21
 - Antennae composed of 11 segments... *Uca* spp. (p. 51, fig. 48b)
21. Number of pleopods
 - Three pairs of pleopods on abdominal somites 2–4 .... *Pinnotheres chamae* (p. 44, fig. 37b)
 - Four pairs of pleopods on abdominal somites 2–5 ....*Pinnotheres ostreum* (p. 44, fig. 38b)

LARVAL DESCRIPTIONS
(Taxonomically arranged)

Category definitions

Synopsis. — Basic life history information for a species. The number of zoeal stages is provided, as are notes on development time in culture and the carapace lengths of larvae (when available). Except where noted, information about the larval lifespan includes the megalopa. Investigators should bear in mind that many factors affect the development time for crustaceans, especially temperature. Thus, the included larval lifespans should be considered approximations.

Distinguishing characteristics. — Characteristic(s) that separate the zoea or megalopa of a given species from all others in the key. This section is intended as a quick reference for investigators as they move through the key.

Description. —

Zoea, general: Characteristics listed in this section are shared by all zoeal stages of a species. When possible, the patterns of melanophores and other chromatophores are included in list form.

Stages: Characteristics found under the specific stage headings can be used to differentiate among zoeal stages of a species. Characteristics listed under a particular stage are present in that zoeal stage and generally in those zoeal stages that follow it.

The zoea I stages of all of the included crabs have sessile eyes and 4 swimming setae on each maxilliped. All zoea II stage crabs have stalked eyes. Because all species share these characteristics, they are not indicated under specific larval descriptions. Additionally, zoeae should be assumed to possess 5 abdominal somites unless otherwise specified.

The phrase "No additional features" under a zoea I heading indicates a zoea with 4 swimming setae on each maxilliped, sessile eyes, and the usual complements of features listed in the "Zoea, general" section for that species.

Euceramus praelongus Stimpson, 1860 (fig. 4)

References: Roberts (1968).

Synopsis: With 2 zoeal stages. Carapace length for zoea I, II and megalopa = 1.02, 1.80 and 2.32 mm, respectively. Rostral spines of zoea I and II = 4.25 and 6.63 mm, respectively.

Zoea distinguishing characteristics: With 2 ventral rows of spinules on rostral spine.
Megalopa distinguishing characteristics: Resembles adult.
Description:

Zoea, general: Carapace triangular with very long rostral and posterior spines. With 2 rows of ventral spinules on rostral spine and 1 row of spinules on posterior spines. No dorsal spine. Spinules on posterior margin of the carapace. Ventro-lateral spines on abdominal somites 4–5. Telson with an anal spine, one pair of lateral spines, and long plumose setae. Chromatophore pattern: Red: (1) base of telson, (2) dorso-laterally on the proximal end of abdominal somites 1–5, (3) maxillipeds.

Zoea I: Telson with 5 pairs of long plumose setae and one pair of small inner setae.

Zoea II: With 9 swimming setae on each maxilliped. Pleopod buds on abdominal somites 2–5. Telson with 6 pairs of long plumose setae.

Megalopa: Rostrum short, broad, and comes to a point. Lateral spine on the margin of the carapace. With 6 abdominal somites. Pleopods on abdominal somites 2–5. Telson with 10 pairs of plumose setae. Pereiopods 1-4 not modified for digging, but pereiopod 5 greatly modified. Chromatophore pattern similar to that of zoeae.

Megalobrachium soriatum (Say, 1818) (fig. 5)

References: Gore (1973).
Synopsis: With 2 zoeal stages. Larval lifespan approximately 20 days at 25°C. Carapace length for zoea I, II and megalopa = 1.00, 1.20 and 1.00 mm, respectively.
Zoea distinguishing characteristics: More than 2 rows of spinules on rostral spine. Very few (1–3) spinules on posterior spines.
Megalopa distinguishing characteristics: Resembles adult.
Description:

Zoea, general: Very long rostral and posterior spines. Rostral spine generally sigmoid shaped, although in some cases may be straight. More than 2 rows of spinules on rostral spine. Posterior spines with very few spinules. Ventral margin of carapace serrate. Lateral spines on abdominal somites 4–5; 5 pairs of plumose setae and one pair of smaller setae on telson. Chromatophore pattern: Red: (1) basipodites on both maxillipeds and the coxa of maxilliped 2, (2) mandible. Additionally, two orange bands with a white band between them near the tip of the rostral spine.

Zoea I: With 3 spinules on posterior spines.

Zoea II: With 10 swimming setae on each maxilliped. With 1 or 2 spinules on posterior spines. Pleopods on abdominal somites 2–4. A single median spine added to the telson in addition to existing setae.

Megalopa: Resembles adult. Entire carapace covered with short spines. Telson with 6 pairs of long plumose setae with shorter setae interspersed between them.

Polyonyx gibbesi Haig, 1956 (fig. 6)

References: Gore (1968).

Synopsis: With 2 zoeal stages. Larval lifespan 30–31 days at 25°C. Carapace length for zoea I and II = 1.20 and 1.70 mm, respectively.

Zoea distinguishing characteristics: More than 2 rows of spinules on rostral spine. Numerous spinules on posterior spines.

Megalopa distinguishing characteristics: Resembles adult.

Description:

Zoea, general: Very long rostral and posterior spines. Rostrum covered with spinules. Numerous spinules on posterior spines. Three pairs of setae on dorsal surface of carapace. Abdominal somites with lateral spines that increase in length closer to telson. Telson with an anal spine and one pair of lateral spines. Chromatophore pattern: (1) surrounding the gut throughout the abdomen on both the dorsal and ventral surface, (2) telson, dorsally between lateral spines, and on tip of telson, (3) around mouthparts.

Zoea I: Telson with 5 pairs of plumose setae and one pair of small inner setae.

Zoea II: Rostral spine about six times carapace length. With 10 swimming setae on each maxilliped. Pleopod buds on abdominal somites 2–5. Telson with 6 pairs of long plumose setae.

Megalopa: Resembles adult. Biramous uropods. Telson with 8 pairs of plumose setae.

Porcellana sigsbeiana A. Milne-Edwards, 1880 (fig. 7)

References: Gore (1971).

Synopsis: With 2 zoeal stages. Larval lifespan for zoea I and II = 9 and 13 days, respectively, at 20°C. Carapace length for zoea I, II and megalopa = 1.12, 1.93 and 2.25 mm, respectively.

Zoea distinguishing characteristics: Hooked spines on basipodites of maxillipeds.

Megalopa distinguishing characteristics: Resembles adult.

Description:

Zoea, general: Very long rostral and posterior spines. Rostral and posterior spines with roughly 4 rows of spinules. Distinct hump on carapace above the midgut. Hooked spines on posterior proximal angle of basipodite of maxillipeds. Lateral spines on abdominal somites 2–5. Telson with one pair of lateral spines. Chromatophore pattern: Red: (1) exopodite of antennae, (2) mandibles, (3) basipodite of maxillipeds, (4) ventral surface of abdominal somites 4 and 5, (5) dorsal surface of abdominal somite 5 and telson. Orange: (1) ventral edge of basipodite of maxillipeds, (2) lateral surface of abdominal somites 2 and 3.

Zoea I: Telson with 5 pairs of long plumose setae and one pair of small inner setae.

Zoea II: With 12 swimming setae on maxilliped 1 and 9 on maxilliped 2. Pleopod buds on abdominal somites 2–5. Telson with 6 pairs of plumose setae.

Megalopa: Resembles adult. Biramous uropods. Variable number of setae on telson, but uropods, exopodites and endopodites all with about 24 setae.

Lepidopa sp. (most likely *L. websteri* Benedict, 1903) (fig. 8)

References: Sandifer & Van Engel (1972b).

Synopsis: With 3 zoeal stages described, hypothesized to have 4 zoeal stages.

Zoea distinguishing characteristics: Large posteriorly curving hooks on abdominal somite 5.

Description:

Zoea, general: Very long rostral and posterior spines without spinules. Rostral spine two to three times carapace length. Small ventro-lateral spines on abdominal somites 3–4. Large posteriorly curving hooks on abdominal somite 5. Telson broad and flat with two pairs of large lateral spines.

Zoea I: With 28–32 setae along posterior margin of telson between the last pair of lateral spines.

Zoea II: With 10 swimming setae on each maxilliped.

Zoea III: With 13–14 swimming setae on each maxilliped; 6 abdominal somites; uropods present on abdominal somite 6.

Emerita talpoida (Say, 1817) (figs. 9a, b)

References: Rees (1959).

Synopsis: With 6 zoeal stages. Development 28 days to the megalopa stage at 30°C. Carapace length for zoeae I–VI and megalopa = 0.57, 0.68, 0.88, 1.13, 1.60, 1.90 and 2.30 mm, respectively.

Zoea distinguishing characteristics: Larva fairly distinctive, but not definitively porcellanid-like or brachyuran-like. Carapace with rostral and postero-lateral spines, but no dorsal spine.

Megalopa distinguishing characteristics: Resembles adult.

Description:

Zoea, general: Carapace with rostral and postero-lateral spines (except zoea I, which lacks postero-lateral spines). Dorsal spine absent. Carapace hemispherical to pear-shaped. Antennae very small. Telson broad, flat, and somewhat concave. With 2 "teeth" on postero-lateral margin of telson.

Zoea I: Rostrum short and triangular. Postero-lateral spines absent. Five abdominal somites (although the first is not differentiated from the abdomen).

Zoea II: Postero-lateral spines present. With 6 swimming setae on each maxilliped.

Zoea III: With 8 swimming setae on each maxilliped. Uropods present.

Zoea IV: With 10 swimming setae on each maxilliped.

Zoea V: With 12 swimming setae on each maxilliped. Endopodite buds on the uropods.

Zoea VI: With 13 or 14 setae on each maxilliped. Uniramous pleopods on abdominal somites 2–5. Endopodites of the uropods about two-thirds the length of the exopodites.

Megalopa: Resembles adult. With 6 abdominal somites. Pleopods on abdominal somites 2–5. Pereiopods 1–3 project anteriorly, while pereiopod 4 projects posteriorly and is modified for digging.

Hypoconcha arcuata Stimpson, 1858 (fig. 10)

References: Kircher (1970).

Synopsis: With 3 zoeal stages.

Zoea distinguishing characteristics: Zoeae shrimp-like in appearance. Carapace without dorsal, rostral, or lateral spines (although 2 antero-lateral grooves are present in zoea III).

Megalopa distinguishing characteristics: Pereiopod 5 is carried over the carapace. Carapace covered with short hairs.

Description:

Zoea, general: Zoeae shrimp-like in appearance. Carapace without dorsal, rostral, or lateral spines. Rostrum short. Both antennae and antennules extend beyond rostrum.

Zoea I: Poorly described. Telson with pronounced medial notch; 7 pairs of processes on the telson, including 5 pairs of long plumose setae.

Zoea II: With 6 swimming setae on each maxilliped. Median notch usually absent from the telson. Telson with 6 pairs of long plumose setae.

Zoea III: Two antero-lateral grooves on carapace. Sharp dorsal ridge along rostrum. Telson rectangular. Pleopods on abdominal somites 2–5. Uropods present. Outer 2 pairs of telson processes are plumose, the remaining 4 pairs are stout and articulating.

Megalopa (referred to as glaucothoë): Anterior medial notch on the carapace. Numerous hairs on the carapace and abdominal somites. Pereiopod 5 carried over carapace. There are 6 abdominal somites with numerous hairs and spines. Telson squarish with 4 pairs of plumose setae along posterior border. Uropods are large and extend past the border of the telson.

Hepatus epheliticus (Linnaeus, 1763) (figs. 11a, b)

References: Costlow & Bookhout (1962a).
Synopsis: With 5 zoeal stages. Larval lifespan is 31–39 days at 25°C and 35 ppt salinity.
Zoea distinguishing characteristics: One row of spines on the distal portion of antennae.
Megalopa distinguishing characteristics: Rostrum appears as a short median knob with lateral knobs.
Description:

Zoea, general: Carapace with dorsal, rostral and lateral spines. Rostral spine fairly short and approximately same length as antennae. One row of spines on the distal portion of the antennae. A small lateral knob and hook on abdominal somites 2 and 3. Postero-lateral spines on abdominal somites 3–5, which overlap following somite. Telson bifurcate with 1 small dorsal and 1 lateral spine on each prong. Chromatophore pattern: (1) postero-lateral to the eyes extending into the base of the rostral spine, (2) dorsal to the heart extending toward lateral spines (the original description cites these chromatophores as being ventral, however, this appears to be a misprint; these chromatophores lie dorsal to the

heart and partially account for the pronounced coloration depicted in the figures), (3) posterior margin of carapace, (4) basipodites of maxillipeds, (5) mandibles and labrum, (6) dorsal surface of abdominal somites 1 and 2, (7) postero-
ventral borders of abdominal somites 2–5.

Zoea I: Telson with 3 + 3 inner spines.

Zoea II: With 6 swimming setae on each maxilliped. Telson with 4 + 4 inner
spines.

Zoea III: Pereiopods may hang below carapace. With 8 swimming setae on
each maxilliped and 6 abdominal somites.

Zoea IV: Pereiopods well developed and always hang below carapace. With
10 swimming setae on each maxilliped. Pleopod buds on abdominal somites 2–6.
Telson with 5 + 5 inner spines.

Zoea V: Pleopod buds well developed. With 11 swimming setae on maxilliped 1 and 12 on maxilliped 2.

Megalopa: Rostrum with a knob on each antero-lateral margin as well as a
downwardly deflected median knob. Antennae with 9 segments. Pleopods on
abdominal somites 2–6. Telson lacks spines. Chromatophore pattern: (1) Postero-lateral edge of rostrum extending into the eyestalks, (2) dorsal surface of the
rostrum lateral to the median line, (3) dorso-median line of the carapace extending posteriorly, (4) antero-lateral margin of carapace, (5) postero-lateral margin
of carapace, (6) abdominal somite 1 dorsal to the gut, (7) dorso-lateral and ventro-lateral surfaces of abdominal somites 2–6, (8) mandibles and labrum, (9)
posterior surfaces of the ischia of the walking legs and chelipeds, and protopo-
dites of the chelipeds.

Persephona mediterranea (Herbst, 1794) (figs. 12a, b)

References: Negreiros-Fransozo et al. (1989).

Synopsis: With 4 zoeal stages. Larval lifespan 35 days at 22–26°C and 34.5 ppt
salinity.

Zoea distinguishing characteristics: Telson fluke-like. Antennae greatly reduced.

Megalopa distinguishing characteristics: Postero-dorsal spine on carapace. No
anteriorly projecting carapace spines. No stiff hairs on dactyl of pereiopod 5.

Description:

Zoea, general: Carapace with dorsal, rostral and lateral spines. Lateral spines
fairly large and generally extend beyond ventral margin of carapace. Antennae
reduced. Postero-lateral knob on abdominal somite 2. Lateral spine on abdomi-

nal somite 3. Telson fluke-like with convex lateral margins and 3 pairs of plumose inner spines.

Zoea I: No additional features.

Zoea II: With 6 swimming setae on each maxilliped.

Zoea III: With 8 swimming setae on each maxilliped.

Zoea IV: Pleopod buds present. With 9 or 10 swimming setae on maxilliped 1 and 10 swimming setae on maxilliped 2.

Megalopa: Postero-dorsal carapace spine. Rostrum rounded. Antennae with 4 segments. No setation on telson.

Anasimus latus Rathbun, 1894 (fig. 13)

References: Sandifer & Van Engel (1972a).

Synopsis: With 2 zoeal stages. Development 21 days to megalopa at 21–25°C.

Zoea distinguishing characteristics: Carapace with dorsal spine, but without lateral or rostral spines.

Distinguishing among the zoeal stages of spider crabs: It is difficult to distinguish among the zoeal stages of most spider crabs (crabs from the genera *Libinia*, *Microphrys*, *Mithrax* and *Stenocionops*). A combination of features must be used to fully differentiate these larvae, and a dissection of mouthparts will almost certainly be required. Investigators are referred to Appendix I for more detailed information on how to differentiate among species.

Megalopa distinguishing characteristics: Carapace with prominent dorsal spine and two anteriorly projecting spines. No stiff hairs on dactyl of pereiopod 5.

Description:

Zoea, general: Carapace with dorsal and ocular spines, but without lateral or rostral spines. Long, anteriorly curved, hook-like extensions on abdominal somite 2. Telson bifurcate with 1 pair of lateral spines and 3 pairs of inner spines.

Zoea I: No additional features.

Zoea II: With 6 swimming setae on each maxilliped. With 6 abdominal somites. Abdominal somites 2–5 with small pleopod buds.

Megalopa: Carapace with prominent dorsal spine and two anteriorly projecting spines. Rostrum terminates in 4 small lobes. Gastric region with a lateral spine on each side and a well-marked median protuberance. Long prominence present on basal article of antennae. With 6 abdominal somites. Telson rounded with 4 terminal setae.

Libinia dubia H. Milne Edwards, 1834 (fig. 14)

References: Sandifer & Van Engel (1971).

Synopsis: With 2 zoeal stages. Larval lifespan 9 days at 25.5–28.5°C. Carapace length of zoea I, II and megalopa = 0.81, 0.97 and 1.16 mm, respectively.

Zoea distinguishing characteristics: Carapace with dorsal and rostral spines, but lacking lateral spines. Rostral spine shorter than antennae.

Distinguishing zoeae of *L. dubia* from those of *L. emarginata*: It is impossible to distinguish *L. dubia* zoeae from *L. emarginata* zoeae. *L. dubia* zoeae, however, generally have 6–8 setae on each side of the postero-lateral margin of the carapace, while *L. emarginata* tend to have 7–10.

Distinguishing among the zoeal stages of spider crabs: See *Anasimus latus*.

Megalopa distinguishing characteristics: Single median cardiac tubercle and no mesogastric tubercle.

Description:

Zoea, general: Large, somewhat rounded carapace with dorsal and rostral spines, but without lateral spines. Dorsal spine fairly long and curved posteriorly. Rostral spine shorter than antennae. Small lateral knobs on abdominal somite 2. Postero-lateral spines on abdominal somites 3–5. Telson bifurcate with a single pair of small lateral spines (1 on each prong) and 3 + 3 inner spines.

Zoea I: With 6–7 setae on each side of postero-lateral margin of the carapace (not visible in the figs.).

Zoea II: With 6 swimming setae on each maxilliped. With 6 abdominal somites. Pleopod buds on abdominal somites 2–5. Postero-lateral spines on abdominal somites 3–5 more pronounced. With 7–8 setae on each side of the postero-lateral margin of the carapace.

Megalopa: Carapace without spines. Small, downwardly directed rostrum. Single median cardiac tubercle. Small ridges on lateral side of the cardiac region and a well marked median cardiac protuberance. With 3 paired protuberances (often small) on lateral carapace region. Abdomen with 6 somites.

Libinia emarginata Leach, 1815 (fig. 15)

References: Johns & Lang (1977).

Synopsis: With 2 zoeal stages. Larval lifespan 14 days at 25°C. Carapace length of zoea I, II and megalopa = 0.78, 0.94 and 1.21 mm, respectively.

Zoea distinguishing characteristics: Carapace with dorsal and rostral spines, but lacking lateral spines. Rostral spine shorter than antennae.

Megalopa distinguishing characteristics: Two cardiac tubercles.

Distinguishing zoeae of *L. dubia* from those of *L. emarginata*: See *Libinia dubia*.

Distinguishing among the zoeal stages of spider crabs: See *Anasimus latus*.

Description:

Zoea, general: Large, somewhat rounded carapace with dorsal and rostral spines, but without lateral spines. Dorsal spine fairly long and curved posteriorly. Rostral spine about half the length of antennae. Small lateral knobs on abdominal somite 2. Small postero-lateral spines on abdominal somites 3–5. Telson bifurcate with a pair of small lateral spines (1 on each prong) and 3 + 3 inner spines.

Zoea I: With 7 setae on each side of the postero-lateral margin of the carapace.

Zoea II: With 6 swimming setae on each maxilliped; 6 abdominal somites. With 8–10 setae each side of the postero-lateral margin of the carapace.

Megalopa: Depressed median line between eyes. With 2 cardiac tubercles and 2 partially connected protuberances along gastric region.

Microphrys bicornutus (Latreille, 1825) (fig. 16)

References: Gore et al. (1982).

Synopsis: With 2 zoeal stages. Larval lifespan 13 days at 25°C. Carapace length for zoea I, II and megalopa = 0.66, 0.82 and 1.19 mm, respectively.

Zoea distinguishing characteristics: Carapace with dorsal and rostral spines, but lacking lateral spines. Rostral spine shorter than antennae.

Distinguishing among the zoeal stages of spider crabs: See *Anasimus latus*.

Megalopa distinguishing characteristics: Rostrum long, directed downwards and strongly hook-like.

Description:

Zoea, general: Carapace with dorsal and rostral spines, but without lateral spines. Rostral spine shorter than antennae. Anteriorly curved lateral knobs on abdominal somite 2. Short postero-lateral spines on abdominal somites 3–5. Telson bifurcate with a small lateral spine on each prong. Overall color, lime green. Melanophore pattern: (1) mandible, (2) basipodites of maxillipeds, (3) pairs on abdominal somites 2–5. Chromatophore pattern: Orange: (1) above the eyes, (2) postero-ventral portion of carapace, (3) base of antennae, (4) basipodite of maxilliped 2, (5) abdominal somites 3–5.

Zoea I: No additional features.

Zoea II: With 6 swimming setae on each maxilliped. Pereiopods well developed and hang below margin of carapace. Pleopod buds on abdominal somites 2–5.

Megalopa: Rostrum long, directed downwards and strongly hook-like. Gastric region of carapace with two pairs of tubercles. Cardiac region of carapace with a single pair of small tubercles. Large, posteriorly directed lobe on the intestinal region of the carapace. Antennae with 6 segments. Melanophore pattern: (1) eye peduncle, (2) rostrum, (3) base of antennae, (4) lateral surface of carapace, (5) dorsal surface of abdominal somites and telson. Chromatophore pattern: Red: (1) pereiopods, (2) ventral surface of abdominal somites 4–6. Orange: (1) base of antennae.

Mithrax pleuracanthus Stimpson, 1871 (fig. 17)

References: Goy et al. (1981).

Synopsis: With 2 zoeal stages. Larval lifespan 9 days at 30°C and 35 ppt salinity. Carapace lengths for zoea I, II and megalopa = 0.96, 1.10 and 1.40 mm, respectively.

Zoea distinguishing characteristics: Carapace with dorsal and rostral spines, but lacking lateral spines. Rostral spine shorter than antennae.

Distinguishing among the zoeal stages of spider crabs: See *Anasimus latus*.

Megalopa distinguishing characteristics: A single median cardiac tubercle and a mesogastric tubercle.

Description:

Zoea, general: Carapace with dorsal and rostral spines, but lacking lateral spines. Rostral spine shorter than antennae, about half the length of antennules. Small lateral knobs on abdominal somite 2. Postero-lateral spines on abdominal somites 3–5. Telson bifurcate with one lateral spine on each prong and 3 + 3 inner spines. Chromatophore pattern: (1) dorsal to the gut, (2) mouthparts, (3) basis of maxilliped 1, coxa and basis of maxilliped 2, (4) abdominal somites 3–5.

Zoea I: With 6 postero-lateral plumose setae on each side of carapace.

Zoea II: With 6 swimming setae on each maxilliped. With 8 postero-lateral plumose setae on each side of carapace.

Megalopa: Pointed rostrum directed obliquely downward. Carapace with numerous rounded tubercles and a total of 5 gastric tubercles (including 1 median cardiac tubercle and 1 mesogastric tubercle). Antennae with 6 segments. "Tooth" on the dactyls of pereiopods 2–4. Pleopods on abdominal somites 2–6. Chromatophore pattern: (1) a pair on dorsal carapace, (2) a pair on postero-lateral carapace, (3) mouthparts, (4) coxae of chelipeds, (5) abdominal somites.

Stenocionops furcatus coelatus (A. Milne-Edwards, 1878) (fig. 18)

References: Laughlin et al. (1984).

Synopsis: With 2 zoeal stages. Carapace lengths for zoea I and II = 1.01 and 1.03 mm, respectively.

Zoea distinguishing characteristics: Carapace with dorsal and rostral spines, but lacking lateral spines. Rostral spine shorter than antennae.

Distinguishing among the zoeal stages of spider crabs: See *Anasimus latus*.

Megalopa distinguishing characteristics: Not described.

Description:

Zoea, general: Carapace with dorsal and rostral spines, but lacking lateral spines. Rostral spine shorter than antennae, about half the length of antennules. Small lateral knobs on abdominal somite 2. Postero-lateral spines on abdominal somites 3–5. Telson bifurcate with one lateral spine on each prong and 3 + 3 inner spines.

Zoea I: No additional features.

Zoea II: With 6 swimming setae on each maxilliped. Pleopods on abdominal somites 2–5.

Cancer irroratus Say, 1817 (figs. 19a, b)

References: Sastry (1977b).

Synopsis: With 5 zoeal stages.

Zoea distinguishing characteristics: Postero-lateral processes on all abdominal somites, but no lateral processes.

Distinguishing zoeae of *C. irroratus* from those of *C. borealis*: A detailed examination of larval appendages is required to differentiate zoeae of *Cancer irroratus* from those of *C. borealis*. Investigators are referred to Sastry (1977a, b) for a more complete treatment of larval characteristics and descriptions of appendages.

Megalopa distinguishing characteristics: Carapace with large dorsal spine. Three stiff hairs on dactyl of pereiopod 5. With arrangement of 15, 15, 13, 11 and 9 plumose setae on pleopods of abdominal somites 2–6, respectively.

Description:

Zoea, general: Carapace with dorsal, rostral, and lateral spines. Dorsal spine curved posteriorly. Antennules about half the length of antennae. Protopodite of antennae has two rows of spines along the distal region. Posterior margins of the abdominal somites bear small processes that overlap next somite (these are knobs in early zoeal stages and become spines in later stages). Telson bifurcate with lateral spines. Chromatophore pattern: (1) mandible, (2) lateral margin of carapace, (3) base of dorsal and lateral spines, (4) margin of third through last abdominal somites.

Zoea I: Protopodite of antennae biramous. Telson with 3 + 3 inner spines.

Zoea II: With 6 swimming setae on each maxilliped.

Zoea III: With 8 swimming setae on each maxilliped. Lateral margin of carapace has 5 spines. With 6 abdominal somites. Telson with 4 + 4 inner spines.

Zoea IV: Pereiopods developed and nearly protrude from beneath carapace. With 10 swimming setae on each maxilliped. Endopodite of antennae half as long as protopodite. Pleopod buds developed on abdominal somites 2–6; buds on somite 6 tiny. Telson with 5 + 5 inner spines.

Zoea V: Pereiopods protrude below carapace. Chelae well developed. With 12 swimming setae on each maxilliped. Endopodite of antennae longer than protopodite and shows evidence of segmentation.

Megalopa: Large dorsal spine on carapace. Antennae with 10 segments. With 6 abdominal somites. There are 3 stiff hairs on dactyl of pereiopod 5. Pleopods on somites 2–6 with 15, 15, 13, 11 and 9 plumose hairs, respectively. Chromatophore pattern: (1) numerous chromatophores on dorsal surface of carapace anterior to dorsal spine, (2) abdominal somites, (3) telson, (4) chelipeds, (5) pereiopods.

Cancer borealis Stimpson, 1859 (figs. 20a, b)

References: Sastry (1977a).

Synopsis: With 5 zoeal stages.

Zoea distinguishing characteristics: Postero-lateral processes on all abdominal somites.

Distinguishing between zoeae of *C. irroratus* and *C. borealis*: See *C. irroratus*.

Megalopa distinguishing characteristics: Carapace with large dorsal spine. With 3 stiff hairs on dactyl of pereiopod 5. Arrangement of plumose setae on the pleopods of abdominal somites 2–6 as 16, 16, 16, 16, 10.

Description:

Zoea, general: Carapace with dorsal, rostral and lateral spines. Dorsal spine curved posteriorly. Antennules about half the length of antennae. Protopodite of antennae has two rows of spines along distal region. Posterior margins of abdominal somites bear small processes that overlap with next somite (these are knobs in early zoeal stages and become spines in later stages). Telson bifurcate with lateral spines. Chromatophore pattern: (1) mandible, (2) lateral margin of carapace, (3) base of dorsal and lateral spines, (4) margin of third through last abdominal somites.

Zoea I: Telson with 3 + 3 inner spines.

Zoea II: With 6 swimming setae on each maxilliped.

Zoea III: With 8 swimming setae on each maxilliped. Lateral margin of the carapace has 5 spines. With 6 abdominal somites. Telson with 4 + 4 inner spines.

Zoea IV: Pereiopods developed and nearly protrude from carapace. With 10 swimming setae on each maxilliped. Endopodite of antennae half as long as protopodite. Pleopod buds developed on abdominal somites 2–6; buds on somite 6 tiny. Telson with 5 + 5 inner spines.

Zoea V: Pereiopods protrude below carapace. Chelae well developed. With 12 swimming setae on each maxilliped. Endopodite of antennae longer than protopodite and shows evidence of segmentation.

Megalopa: Large dorsal spine on carapace. Antennae with 11 segments. Three stiff hairs project from dactyl of pereiopod 5. Six abdominal somites. Pleopods on abdominal somites 2–6 with 16, 16, 16, 16 and 10 plumose hairs, respectively. Chromatophore pattern: (1) numerous chromatophores on dorsal surface of carapace anterior to dorsal spine, (2) telson, (3) chelipeds, (4) pereiopods.

Ovalipes ocellatus (Herbst, 1799) (figs. 21a, b)

References: Costlow & Bookhout (1966b).

Synopsis: With 5 zoeal stages. Larval lifespan 18 days at 25°C and 30 ppt salinity, 26 days at 20°C and 30 ppt salinity.

Zoea distinguishing characteristics: Lateral knob on abdominal somite 2. Postero-lateral spines on abdominal somites 3–5. Lateral hook on at least abdominal somite 5 and sometimes on abdominal somites 3–4.

Megalopa distinguishing characteristics: Dactyl of pereiopod 5 flattened and with 3 stiff hairs.

Description:

Zoea, general: Carapace with dorsal, rostral and lateral spines. Hooks or knobs on all abdominal somites except on somite 1. Lateral knob on abdominal somite 2. Postero-lateral spines on abdominal somites 3–5. Lateral hooks and postero-lateral spines on somites 3–5 (however, see below for exceptions in zoea IV and V). Telson with 2 pairs of lateral spines, but lacking dorsal spines. Chromatophore pattern: (1) a pair medial and dorsal to the eyes, (2) a pair anterior and slightly lateral to base of dorsal spine, (3) posterior to the eyes, (4) dorsal to the heart, (5) one dorsal and one ventral to lateral spine, (6) mandibles and labrum, (7) abdominal somite 1 dorsal to the gut, (8) postero-ventral surface of abdominal somites 2–5, (9) basipodite of maxillipeds.

Zoea I: Telson with 3 + 3 inner spines.

Zoea II: With 7 swimming setae on each maxilliped.

Zoea III: With 8 swimming setae on maxilliped 1 and 10 on maxilliped 2. With 6 abdominal somites. Tiny pleopod buds on all abdominal somites. Telson with 4 + 4 inner spines.

Zoea IV: Pereiopods visible below carapace. With 10 swimming setae on maxilliped 1 and 12 on maxilliped 2. Pleopod buds are enlarged on abdominal somites 2–6. The lateral projection on abdominal somite 4 (not the postero-lateral spine) is absent. Telson with 5 + 5 inner spines.

Zoea V: With 13 swimming setae on maxilliped 1 and 15 on maxilliped 2. Pereiopods well developed.

Megalopa: Rostrum ends in a single, downwardly deflected median process. Small anteriorly projecting hook on basi-ischium of each chela. The dactyl of pereiopod 5 flattened and with 3 stiff hairs. Pleopods on abdominal somites 2–6. Chromatophore pattern: (1) dorsal surface of eyestalks, (2) a pair on dorsal surface of rostrum, (3) posterior to each eyestalk, (4) posterior margin of carapace, (5) 3 along the median line of dorsal surface of carapace, (6) dorsal to the gut on abdominal somites 1 and 2, (7) postero-lateral borders of abdominal somites 2–6.

Arenaeus cribrarius (Lamarck, 1818) (fig. 22)

References: Sandifer (1972).

Synopsis: Only zoea I described.

Zoea distinguishing characteristics: Lateral knobs or hooks on abdominal somites 2 and 3. Small, bifid postero-lateral spines on abdominal somites 3–5.

Description:

Zoea, general: Carapace with dorsal, rostral and lateral spines. Dorsal spine curves strongly posteriorly. Rostral spine curves slightly posteriorly near its tip. Small lateral knobs on abdominal somites 2 and 3. Small bifid postero-lateral spines on abdominal somites 3–5. Telson bifurcate with 1 pair of dorsal spines and 2 pairs of lateral spines (one large, the other small).

Zoea I: Telson with 3 + 3 inner spines.

Callinectes sapidus Rathbun, 1896 (figs. 23a, b, c)

References: Costlow & Bookhout (1959).

Synopsis: With 7 zoeal stages (a zoea VIII stage was sometimes seen, but never successfully metamorphosed into a megalopa). Larval lifespan 37–55 days at 26.7°C and 25 ppt salinity.

Zoea distinguishing characteristics: Lateral knob or hook on abdominal somites 2 and 3. Postero-lateral spines on abdominal somites 3–5. Two rows of spines on distal portion of the antennae. No chromatophore on dorsal spine.

Distinguishing among zoeae of the subfamily Portuninae: It is difficult to distinguish among zoeae of the Portuninae (*C. sapidus*, *C. similis* and *Portunus spinicarpus*), especially between *C. sapidus* and *C. similis*, although the zoeae of these species can generally be differentiated by their setation. Investigators are referred to Bookhout & Costlow (1977) and to Appendix II for more detailed descriptions of zoeal setation.

Megalopa distinguishing characteristics: Antennae with 11 segments. Megalopae of *C. sapidus* and *C. similis* are very similar, although megalopae of *C. sapidus* are often much larger than those of *C. similis*. Additionally, there are differences in setation (see Appendix II), and *C. sapidus* has 5 hairs on the dactyl of pereiopod 5, while *C. similis* has 8.

Description:

Zoea, general: Carapace with dorsal, rostral and lateral spines. Rostral spine short relative to carapace length and does not protrude beyond the maxillipeds. Two rows of spines on the distal half of antennal protopodite. Small lateral knob on abdominal somite 2. Small lateral hook on somite 3. Postero-lateral spines on abdominal somites 3–5. Telson bifurcate with each prong having a lateral and (smaller) dorsal spine. Chromatophore pattern: (1) between the eyes, (2) posterior to eyes, dorso-lateral to anterior portion of the gut, (3) posterior carapace, dorsal to the gut, (4) below dorsal spine, (5) mandible, (6) basipodite of maxilliped 1, (7) median portion of abdominal somite 1 dorsal to the gut, (8) margin of abdominal somites 3–5 (and 6 when added).

Zoea I: Telson with 3 + 3 inner spines.

Zoea II: With 6 swimming setae on each maxilliped. Telson with 4 + 4 inner spines.

Zoea III: With 8 swimming setae on each maxilliped. With 6 abdominal somites.

Zoea IV: With 9 swimming setae on each maxilliped. Endopodite bud may be visible on the antennae, but is very small.

Zoea V: With 9 swimming setae on maxilliped 1 and 11 on maxilliped 2. Developing endopodite bud on antennae plainly visible.

Zoea VI: With 11 swimming setae on maxilliped 1 and 12 on maxilliped 2. Pleopod buds on abdominal somites 2–6. Single inner spine added between the existing 4 pairs of spines on telson, for a total of 9 spines.

Zoea VII: Pereiopods hang below carapace. With 14 swimming setae on maxilliped 1 and 13 on maxilliped 2.

Zoea VIII (when present): With 12 swimming setae on maxilliped 1 and 14 on maxilliped 2. Hairs on pleopod buds. Telson with a total of 10 spines on the inner margin.

Megalopa: Rostrum ends in long pointed spine. Antennae composed of 11 segments. Spine on lateral surface of the basi-ischium of each cheliped. The dactyl of pereiopod 5 bears 5 spines. A pair of spines projects from posterior edge of carapace. Large postero-lateral spines on abdominal somite 5 that project past abdominal somite 6.

Callinectes similis Williams, 1966 (figs. 24a, b, c)

References: Bookhout & Costlow (1977).

Synopsis: With 8 zoeal stages.

Zoea distinguishing characteristics: A single lateral knob or hook on abdominal somites 2 and 3. Postero-lateral spines on abdominal somites 3–5. Two rows of the spines on the distal portion of the antennae. No chromatophore on dorsal spine.

Distinguishing between zoeae of the subfamily Portuninae: See *C. sapidus*.

Megalopa distinguishing characteristics: Antennae with 11 segments. See *C. sapidus*.

Description:

  Zoea, general: Carapace with dorsal, rostral and lateral spines. Rostral spine short relative to carapace length and does not protrude beyond the maxillipeds. Two rows of spines on the distal half of antennal protopodite. Small lateral knob on abdominal somite 2. Small lateral hook on abdominal somite 3. Postero-lateral spines on abdominal somites 3–5. Telson bifurcate with each prong having a lateral and (smaller) dorsal spine. Chromatophore pattern: (1) posterior to eyes, (2) base of antennae and antennules, (3) base of rostrum, (4) carapace antero-dorsal to the gut, (5) two on the carapace dorsal to the gut, (6) single chromatophore on basipodite of maxillipeds, (7) mandibles and labrum, (8) around the gut on abdominal somites 1 and 2, (9) a pair on postero-lateral borders of abdominal somites 3–5.

Zoea I: Telson with 3 + 3 inner spines.

Zoea II: With 6 swimming setae on each maxilliped. Telson with 4 + 4 inner spines.

Zoea III: With 8 swimming setae on each maxilliped and 6 abdominal somites.

Zoea IV: With 9 swimming setae on maxilliped 1 and 10 on maxilliped 2. Single inner spine added between the existing 4 pairs of spines on telson, for a total of 9 spines.

Zoea V: With 11 swimming setae on maxilliped 1 and 12 on maxilliped 2. Developing endopodite bud on antennae plainly visible. Telson with 5 + 5 inner spines.

Zoea VI: With 12 swimming setae on maxilliped 1 and 14 on maxilliped 2. Pleopod buds on abdominal somites 2–6.

Zoea VII: Pereiopods hang below carapace. With 14 swimming setae on maxilliped 1 and 15 on maxilliped 2. Single inner spine added between existing 5 pairs of spines on telson, for a total of 11 spines.

Zoea VIII: With 15 swimming setae on maxilliped 1 and 17 on maxilliped 2.

Megalopa: Rostrum ends in long pointed spine. Antennae composed of 11 segments. Spine on lateral surface of basi-ischium of each cheliped. The dactyl of pereiopod 5 bears 8 spines. A pair of spines project from the posterior edge of the carapace. Large postero-lateral spines on abdominal somite 5 that project past abdominal somite 6.

Portunus spinicarpus (Stimpson, 1871) (figs. 25a, b, c)

References: Bookhout & Costlow (1974).

Synopsis: With 7 zoeal stages. Larval lifespan 49 days at 25°C and 35 ppt salinity.

Zoea distinguishing characteristics: Single lateral knob or hook on abdominal somites 2 and 3. Postero-lateral spines on abdominal somites 3–5. Two rows of spines on the distal portion of antennae. Single chromatophore on dorsal spine.

Distinguishing among zoeae of the subfamily Portuninae: See *Callinectes sapidus*.

Megalopa distinguishing characteristics: Antennae with 10 segments.

Description:

Zoea, general: Carapace with dorsal, rostral and lateral spines. Rostral spine short relative to carapace length and does not protrude beyond maxillipeds. Two rows of spines on distal portion of antennae. Small lateral knob on abdominal somite 2. Small lateral hook on abdominal somite 3. Postero-lateral spines on abdominal somites 3-5 that overlap following somite. Telson bifurcate with each prong bearing a lateral and (smaller) dorsal spine. Chromatophore pattern: (1) single chromatophore between the eyes, (2) single chromatophore on dorsal spine, (3) posterior portion of carapace behind lateral spines, (4) ventral to lateral spine, (5) antero-dorsal to the gut, (6) two dorsal to the gut, (7) mandible and labrum, (8) basipodite of maxilliped 1, (9) a pair each on the ventral surface of abdominal somites 2–6.

Zoea I: Telson with 3 + 3 inner spines.

Zoea II: With 6 swimming setae on each maxilliped. Telson with 4 + 4 inner spines.

Zoea III: With 8 swimming setae on each maxilliped. With 6 abdominal somites.

Zoea IV: With 10 swimming setae on each maxilliped. Endopodite bud may be visible on antennae, but is very small.

Zoea V: With 10 swimming setae on maxilliped 1 and 11 on maxilliped 2. The developing endopodite bud on antennae plainly visible.

Zoea VI: With 13 swimming setae on maxilliped 1 and 14 on maxilliped 2. Pleopod buds on abdominal somites 2–6. A single inner spine added between existing 4 pairs of spines on telson, for a total of 9 spines.

Zoea VII: Pereiopods hang below carapace. With 14 swimming setae on maxilliped 1 and 15 on maxilliped 2.

Megalopa: Rostrum ends in a single forwardly directed spine. Carapace lacking dorsal or lateral spines. Antenna composed of 10 segments. Spine on lateral surface of basi-ischium of each cheliped. Spine on the ventral side of basipodite of pereiopod 2. A pair of spines on posterior edge of carapace that project beyond pereiopod 5. Dactyl of pereiopod 5 flattened and is bordered by long hooked setae. Large postero-lateral spines on abdominal somite 5 that project past abdominal somite 6.

Pseudomedaeus agassizii (A. Milne-Edwards, 1880) (figs. 26a, b)

References: Costlow & Bookhout (1968).

Synopsis: With 4 zoeal stages. Larval lifespan 33–37 days at 25°C and 35 ppt salinity.

Zoea distinguishing characteristics: Secondary rostral spines in zoeal stages II–IV (not present in zoea I). Zoea I has more than 2 rows of spines on antenna.

Megalopa distinguishing characteristics: The megalopa is very similar to that of *Panopeus herbstii*, but can be differentiated by examining the antennules. There are 3 large aesthetascs on each side of the base of the segmented flagellum in *P. herbstii* that are lacking in *Pseudomedaeus agassizii*. In addition, there are two setae between the rostrum and each lateral horn in *P. agassizii* that are lacking in *P. herbstii*.

Description:

Zoea, general: Carapace with dorsal, rostral and lateral spines. More than 2 rows of spines on antennae. Small lateral knob on abdominal somite 2. Small lateral hook on abdominal somite 3. Long postero-lateral spines on abdominal somites 3–5 that overlap following somite. Telson bifurcate with two pairs of lateral spines, one small, one minute, and one pair of small dorsal spines. Chromatophore pattern: (1) a pair between the eyes, (2) slightly dorsal and posterior to eyes, (3) posterior to eyes on ventral margin of carapace, (4) margin of carapace posterior to lateral spine, (5) margin of carapace ventral to lateral spine, (6) basi-ischium of maxillipeds, (7) labrum and mandible, (8) base of antennae and antennules, (9) abdominal somite 1 dorsal to the gut, (10) postero-lateral borders of abdominal somites 2-6.

Zoea I: Telson with 3 + 3 inner spines.

Zoea II: A small secondary spine on each side of rostrum (i.e., secondary rostral spines). With 6 swimming setae on maxilliped 1 and 7 on maxilliped 2.

Zoea III: Pereiopods hang below carapace. With 8 swimming setae on maxilliped 1 and 9 on maxilliped 2. With 6 abdominal somites. Pleopod buds on abdominal somites 2–6. Telson with 4 + 4 inner spines.

Zoea IV: With 9 swimming setae on maxilliped 1 and 10 on maxilliped 2. A single spine added between the 4 existing pairs of spines on telson, for a total of 9 spines.

Megalopa: The rostrum ends in a downwardly deflected median spine and two stout lateral spines. Three large aesthetascs on each side of the base of the segmented flagellum. Two setae between rostrum and each lateral horn. Antennae with 9 segments. The basi-ischium of each cheliped bears a posteriorly curved hook. Pleopods on abdominal somites 2–6. Chromatophore pattern: (1) median and posterior to eyestalks extending into the rostrum, (2) dorsal surface of eyestalks, (3) postero-lateral surface of carapace, (4) basi-ischium of chelae, (5) dactyls of chelae, (6) abdominal somite 1 dorsal to the gut, extending into abdominal somite 2, (7) postero-lateral borders of abdominal somites 2–6.

Rhithropanopeus harrisii (Gould, 1841) (figs. 27a, b)

References: Chamberlain (1962), Hood (1962).

Synopsis: With 4 zoeal stages. Fifteen days to megalopa at 27°C.

Zoea distinguishing characteristics: Very long rostral spine and antennae relative to carapace length. Pair of long postero-lateral spines on abdominal somite 5. No terminal spines on antennae.

Megalopa distinguishing characteristics: Poorly described.

Description:

    Zoea, general: Carapace with dorsal, rostral and lateral spines. Rostral spine and antennae very long relative to carapace length. Antennae lack spines. Large postero-lateral spine on abdominal somite 5. All other somites lack postero-lateral spines. Abdominal somite 2 has short mid-lateral projections. Telson bifurcate with small dorsal spines.

    Zoea I: Telson with 3 + 3 inner spines.

    Zoea II: With 6 swimming setae on maxilliped 1 and 7 on maxilliped 2.

    Zoea III: With 6 abdominal somites. Pleopod buds distinct. With 8 swimming setae on maxilliped 1 and 9 on maxilliped 2. Telson with 4 + 4 inner spines.

    Zoea IV: With 9 swimming setae on maxilliped 1 and 11 on maxilliped 2. Endopodite bud of antennae visible. Pleopods on somites 2–5 much larger and biramous.

    Megalopa: Rostrum short and notched. Antennae with 9 segments. With 6 abdominal somites. Biramous pleopods on somites 2–5. Uniramous pleopods on somite 6. Telson not described.

Micropanope sculptipes Stimpson, 1871 (figs. 28a, b)

References: Andryszak & Gore (1981).

Synopsis: With 4 zoeal stages. Lifespan 25 days to megalopa at 25°C and 36 ppt salinity. Carapace length for zoeae I–IV and megalopa = 0.47, 0.61, 0.76, 0.84 and 1.25 mm, respectively.

Zoea distinguishing characteristics: Telson with one small pair of lateral spines.

Megalopa distinguishing characteristics: With 4 plumose setae between rostrum and each lateral horn.

Description:

    Zoea, general: Carapace with dorsal, rostral and lateral spines. Antenna approximately the same length as rostral spine, with the distal portion of the protopodite covered with small spinules. Antennule less than half the length of antenna. Postero-lateral spines on abdominal somites 3-5. Small lateral hook-like projections on somites 2 and 3. Telson bifurcate with one small lateral spine and one small dorsal spine on each prong.

Zoea I: Telson with 3 + 3 inner spines.

Zoea II: With 6 swimming setae on each maxilliped.

Zoea III: Secondary rostral protuberances (referred to as pre-orbital protuberances in the description by Andryszak & Gore, 1981). With 8 swimming setae on maxilliped 1 and 9 on maxilliped 2. Pleopod buds on abdominal somites 2–6. Telson with 4 + 4 inner spines.

Zoea IV: Pereiopods well developed and hang below carapace. With 9 swimming setae on maxilliped 1 and 11 on maxilliped 2. Telson with a total of 9 or 10 inner spines.

Megalopa: Rostrum triangular and sharply rounded with prominent antero-lateral horns. With 4 plumose setae between rostrum and each lateral horn. Spine on basi-ischium of each cheliped. Antennae with 8 segments. Telson somewhat squarish with 2 pairs of dorsal setae and a total of 9 setae along posterior margin.

Eurypanopeus depressus (Smith, 1869) (figs. 29a, b)

References: Costlow & Bookhout (1961a).

Synopsis: With 4 zoeal stages.

Zoea distinguishing characteristics: Small terminal spines on antennal protopodite.

Megalopa distinguishing characteristics: Spine on basi-ischium of each cheliped. No posterior carapace spines, rostral horns, or stiff hairs on dactyl of pereiopod 5. Seven terminal setae on telson.

Description:

Zoea, general: Carapace with dorsal, rostral and lateral spines. Dorsal spine curves slightly posteriorly. Antennae approximately same length as rostral spine. Antennae relatively long with small spines near the terminal end. Postero-lateral spines on abdominal somites 3–5. Small lateral hook-like projections on abdominal somites 2 and 3. Telson bifurcate without lateral spines, but with a small dorsal spine on each prong. Chromatophore pattern: (1) dorsal and median to each eye, (2) posterior and slightly ventral to each eye, (3) dorsal to the heart, (4) mandibles and labrum, (5) antennules, (6) basipodite of maxillipeds, (7) postero-lateral margins of carapace, (8) abdominal somite 1 dorsal to the gut, (9) postero-ventral surface of abdominal somites 2–5, (10) telson. *Note*: In figs. 29a and b, the dorsal spines of the stage III and IV zoeae appear to be relatively short compared to those of stage I and II zoeae; it is possible that this is due to laboratory rearing conditions.

Zoea I: Telson with 3 + 3 inner spines.

Zoea II: Slight swelling at basal portion of antennae. With 6 swimming setae on maxilliped 1 and 7 swimming setae on maxilliped 2.

Zoea III: Pereiopods extend below carapace. With 8 swimming setae on maxilliped 1 and 9 swimming setae on maxilliped 2. With 6 abdominal somites. Small pleopod buds on ventral side of abdominal somites 1–6. Telson with 4 + 4 inner spines.

Zoea IV: With 9 swimming setae on maxilliped 1 and 11 swimming setae on maxilliped 2. Total of 9 inner spines on telson.

Megalopa: Rostrum terminates in a blunt median process, which is directed downwards (but much less pronounced than in *P. herbstii*). Pleopods on abdominal somites 2–6. With 3 small setae on telson and 2 additional pairs of longer setae on each side (a total of 7 setae). Spine on the basi-ischium of each cheliped. Antenna with 9 segments. Chromatophore pattern: (1) median to eyes extending into rostrum, (2) posterior to eyes, (3) postero-dorsal surface of carapace, (4) lateral and posterior margin of carapace, (5) abdominal somite 1 dorsal to the gut, (6) postero-lateral borders of abdominal somites 2–6.

Dyspanopeus sayi (Smith, 1869) (fig. 30)

References: Chamberlain (1961), Sandifer (1972).

Synopsis: With 4 zoeal stages.

Zoea distinguishing characteristics: No spines on antennal protopodite.

Megalopa distinguishing characteristics: Not described.

Description:

Zoea, general: Carapace with dorsal, rostral and lateral spines. Antennae approximately the same length as rostral spine and lack terminal spines on protopodite. Postero-lateral spines on abdominal somites 3–5. Small lateral hook-like projections on abdominal somites 2 and 3. Telson bifurcate without lateral spines, but with one small dorsal spine on each prong.

Zoea I: Telson with 3 + 3 inner spines.

Zoea II: With 6 swimming setae on maxilliped 1 and 7 on maxilliped 2. Telson with 4 + 4 inner spines.

Zoea III: With 7–8 swimming setae on maxilliped 1 and 8–9 on maxilliped 2. With 6 abdominal somites. Pleopod buds present on abdominal somites 2–6.

Zoea IV: Pereiopods well developed and hang below carapace. With 9–10 swimming setae on each maxilliped.

Panopeus herbstii H. Milne Edwards, 1834 (figs. 31a, b)

References: Costlow & Bookhout (1961b).

Synopsis: With 4 zoeal stages.

Zoea distinguishing characteristics: Telson with two pairs of lateral spines, one large and one small.

Megalopa distinguishing characteristics: With 1 seta between rostrum and each rostral horn. With 3 large aesthetascs on each side of the base of antenna flagellum. See also description of *Pseudomedaeus agassizii* megalopa.

Description:

Zoea, general: Carapace with dorsal, rostral and lateral spines. Dorsal spine curves slightly posteriorly. Rostral spine slightly longer than antennae. Antennae relatively long with small spines near the terminal end. Postero-lateral spines on abdominal somites 3–5. Small hook-like projection on abdominal somite 2. Telson bifurcate with lateral spines. Chromatophore pattern: (1) dorsal and median to each eye, (2) posterior and slightly ventral to each eye, (3) dorsal to the heart posterior to dorsal spine, (4) postero-lateral margins of carapace, (5) mandibles and labrum, (6) abdominal somite 1 dorsal to the gut, (7) postero-ventral surface of abdominal somites 2–5 (and postero-lateral surface of somite 6 when added), (8) basi-ischium of maxillipeds 1 and 2.

Zoea I: Telson with 3 + 3 inner spines.

Zoea II: Slight swelling on basal portion of antennae. With 6 swimming setae on maxilliped 1 and 7 swimming setae on maxilliped 2.

Zoea III: Pereiopods extend below carapace. With 8 swimming setae on maxilliped 1 and 9 swimming setae on maxilliped 2. With 6 abdominal somites. Small pleopod buds on ventral side of somites 2–6. Telson with 4 + 4 inner spines.

Zoea IV: With 10 swimming setae on maxilliped 1 and 11 swimming setae on maxilliped 2. Telson with a total of 9 inner spines.

Megalopa: Rostrum ends in a median blunt, hump-like process. Prominent lateral horns. One seta between rostrum and each lateral horn. Antenna with 9 segments. Antennule with 3 large aesthetascs on each side of flagellum. Hook on basi-ischium of each cheliped. Three stiff setae on dactyl of pereiopod 5. Pleopods on abdominal somites 2–6. With 3–6 spines on caudal margin of telson. Chromatophore pattern: (1) median to the eyes extending into rostrum, (2) posterior to the eyes, (3) postero-dorsal surface of carapace, (4) lateral and posterior margins of carapace, (5) basi-ischium of chelae and dactyls of chelae, (6) abdominal somite 1 dorsal to the gut, (7) postero-lateral borders of somites 2–6.

Hexapanopeus angustifrons (Benedict & Rathbun, 1891) (figs. 32a, b)

References: Costlow & Bookhout (1966a).

Synopsis: With 4 zoeal stages. Larval lifespan 21–37 days, depending on temperature.

Zoea distinguishing characteristics: No dorsal or lateral spines on telson.

Megalopa distinguishing characteristics: Two pronounced rostral horns and depressed rostrum. Spine on basi-ischium of each cheliped.

Description:

Zoea, general: Carapace with dorsal, rostral and lateral spines. Dorsal spine curves slightly posteriorly. Antennae relatively long, smooth and approximately equal in length to rostral spine. Postero-lateral spines on abdominal somites 3–5. Small projections on somite 2. Telson bifurcate without lateral or dorsal spines. Chromatophore pattern: (1) dorsal and median to each eye, (2) posterior and slightly ventral to each eye, (3) mandibles and labrum, (4) antennules, (5) basipodite of maxillipeds 1 and 2, (6) dorsal to the heart, (7) postero-lateral margins of carapace, (8) abdominal somite 1 dorsal to the gut, (9) postero-ventral surface of abdominal somites 2–5 (and postero-lateral surface of somite 6 when added).

Zoea I: Telson with 3 + 3 inner spines.

Zoea II: With 6 swimming setae on maxilliped 1 and 7 swimming setae on maxilliped 2.

Zoea III: Pereiopods extend below carapace. With 8 swimming setae on maxilliped 1 and 9 swimming setae on maxilliped 2. With 6 abdominal somites. Small pleopod buds on somites 2–6. Telson with 4 + 4 inner spines.

Zoea IV: With 9 swimming setae on maxilliped 1 and 10 swimming setae on maxilliped 2. Telson with a total of 9 inner spines.

Megalopa: Rostrum depressed along median line and terminates in two prominent lateral horns. With 6 abdominal somites. Pleopods on somites 2–6. Hooks on basi-ischium of each cheliped. Antenna with 9 segments. Chromatophore pattern: (1) median to the eyes extending into rostral horns, (2) posterior to eyes, (3) dorsal surface of eyestalks, (4) a pair on lateral margin of carapace, (5) a pair on posterior margin of carapace, (6) first abdominal somite, dorsal to the gut extending into carapace, (7) postero-lateral borders of abdominal somites 2–6.

Pilumnus dasypodus Kingsley, 1879 (figs. 33a, b)

References: Sandifer (1974), Bookhout & Costlow (1979).

Synopsis: With 4 zoeal stages. Carapace lengths for zoeae I–IV and megalopa = 0.44, 0.48, 0.65, 0.77 and 1.03, respectively.

Zoea distinguishing characteristics: Rostral spine shorter than antennae.

Megalopa distinguishing characteristics: No hairs on dactyl of pereiopod 5. No terminal setae on telson.

Description:

Zoea, general: Carapace with dorsal, rostral and lateral spines. Dorsal spine curves posteriorly. Rostral spine significantly shorter than antennae. Small spines near the terminal end of antennae. Postero-lateral spines on abdominal somites 3–5. Small horn-like projection on somite 2. Telson bifurcate with a small dorsal spine and pair of lateral spines. Telson with 3 + 3 inner spines. Chromatophore pattern: (1) base of dorsal spine, (2) on carapace posterior to lateral spines, (3) lateral side of carapace above the gut in the region of the heart, (4) mandible and labrum, (5) dorsal to eyes, (6) base of antennules, (7) base of maxillipeds 1 and 2, (8) ventral side of abdominal somites 2–5, (9) base of pereiopod buds in later zoeal stages.

Zoea I: No additional features.

Zoea II: With 6 swimming setae on each maxilliped.

Zoea III: With 8 swimming setae on each maxilliped. With 6 abdominal somites. Small pleopod buds on ventral side of abdominal somites 2–6.

Zoea IV: Pereiopods developed. With 10 swimming setae on each maxilliped. Biramous pleopods on abdominal somites 2–5. Uniramous pleopods on somite 6.

Megalopa: Reduced downward pointed rostrum without lateral knobs. Pleopods on somites 2–6. Antennae with 9 segments. Two hook-like spines on inner ramus of pleopods on abdominal somites 2–5. Telson broader than long. No terminal setae on telson.

Menippe mercenaria (Say, 1818) (figs. 34a, b)

References: Porter (1960).

Synopsis: With 5 zoeal stages (6 zoeal stages described from laboratory culture, but zoea VI believed to be an anomaly). Larval lifespan 27 days to megalopa.

Zoea distinguishing characteristics: Antennae about half the length of rostral spine.

Megalopa distinguishing characteristics: Not described.

Description:

Zoea, general: Carapace with dorsal, rostral and lateral spines. Antennae about half the length of rostral spine. Anteriorly directed lateral spines on abdominal somite 2. Posteriorly directed lateral spines on abdominal somite 3. Postero-lateral spines on abdominal somites 3–5. Telson bifurcate with a small dorsal spine on each prong. Chromatophore pattern: (1) inter-orbital, (2) dorsal to the heart, (3) anterior surface of dorsal spine, (4) postero-ventral margin of carapace, (5) mandible and labrum, (6) base of antennules, (7) base of antennae, (8) basipodite of maxillipeds 1 and 2, (9) abdominal somite 1 dorsal to the gut, (10) ventro-laterally on abdominal somites 2–5, (11) telson.

Zoea I: Telson with 3 + 3 inner spines.

Zoea II: With 6 swimming setae on each maxilliped. A pair of spines on postero-ventral surface of abdominal somites 3–5.

Zoea III: With 8 swimming setae on each maxilliped. With 6 abdominal somites. Telson with 4 + 4 inner spines.

Zoea IV: With 10 swimming setae on each maxilliped. Pleopod buds on abdominal somites 2–6. Telson with 5 + 5 inner spines.

Zoea V: Pereiopods protrude beneath carapace. With 11 swimming setae on maxilliped 1 and 12 on maxilliped 2. Pleopods well developed.

Zoea VI (when present): Very similar to zoea V.

Dissodactylus crinitichelis Moreira, 1901 (figs. 35a, b)

References: Pohle & Telford (1981).

Synopsis: With 3 zoeal stages. Larval lifespan 7 days at 28°C and 33 ppt salinity. Carapace lengths for zoea I–III and megalopa = 0.44, 0.52, 0.58 and 0.58 mm, respectively.

Zoea distinguishing characteristics: Rostral spine as long as, or longer than, dorsal spine.

Distinguishing between zoeae of *Dissodactylus crinitichelis* and *D. mellitae*: Given the limited information available concerning larval development in *Dissodactylus mellitae*, it is currently impossible to distinguish between larvae of *D. crinitichelis* and *D. mellitae*.

Megalopa distinguishing characteristics: Carapace without spines and subovate in profile.

Description:

Zoea, general: Carapace with dorsal, rostral and lateral spines. Lateral spines are long and extend well below ventral margin of carapace. Rostral spine as long as, or longer than, dorsal spine. Antennae are very short with two rows of spinules on the protopodite. Lateral spines on abdominal somites 2 and 3. Telson bifurcate without dorsal or lateral spines and with 3 + 3 inner spines. Melano-phore pattern: (1) under base of rostral spine extending into antennules, (2) between the eyes, (3) posterior to base of dorsal spine, (4) lateral to the gut on carapace, (5) posterior ventro-lateral margin of carapace extending into lateral spines, (6) basipodite of maxilliped 1, (7) mandible, (8) a pair dorso-lateral to the stomach, (9) a pair each on ventral side of abdominal somites 2, 4 and 5, (10) base of antennae. Chromatophore pattern: Red: (1) coxa of maxilliped 2, (2) posterior to eyes, (3) lateral sides of abdominal somite 1.

Zoea I: No additional features.

Zoea II: With 6 swimming setae on each maxilliped. Lateral spines on abdominal somites 2–5 are long and overlap following somite. Pleopod buds on abdominal somites 2–5.

Zoea III: With 8 swimming setae on each maxilliped.

Megalopa: Carapace without spines and sub-ovate in profile. Rostrum ends in a reduced point. Antennae relatively small with a globular base. Telson rounded and lacks marginal setae. Melanophore pattern: similar to zoeae, with the following adjustments: (1) between the eyes, (2) posterior dorso-lateral carapace, (3) mid-posterior and ventro-lateral margins of carapace, (4) coxae of chelipeds, (5) all segments of pereiopods 2–5 except the dactyls.

Dissodactylus mellitae (Rathbun, 1900) (fig. 36)

References: Sandifer (1972).
Synopsis: Only zoea I has been described.
Zoea distinguishing characteristics: Rostral spine as long as, or longer than, dorsal spine.
Distinguishing between zoeae of *Dissodactylus crinitichelis* and *D. mellitae*: See *D. crinitichelis*.
Description:

Zoea, general: Carapace with dorsal, rostral, and lateral spines. Lateral spines are long and curved ventrally. Rostral spine longer than dorsal and lateral spines. Small knobs on abdominal somites 2 and 3. Telson bifurcate without dorsal and lateral spines and with a fairly straight margin. Antennae slightly longer than antennules.

Zoea I: Telson with 3 + 3 inner spines.

Pinnotheres chamae Roberts, 1975 (figs. 37a, b)

References: Roberts (1975).

Synopsis: With 3 zoeal stages. Larval lifespans of zoeae I–III, and megalopa = 3.2, 3.7, 3.7 and 1.5 days, respectively, at 25.3°C.

Zoea distinguishing characteristics: Carapace without dorsal or lateral spines. Rostral spine greatly reduced.

Distinguishing between zoeae of *Pinnotheres chamae* and *P. ostreum*: It is impossible to distinguish between the zoea I of *Pinnotheres chamae* and *P. ostreum*. Later zoeal stages may be distinguished by the setation of the dorsal surface of abdominal somite 1. Zoeae II and III of *P. chamae* have 3 and 5 setae, respectively, whereas zoeae of *P. ostreum* have no setae. Additionally, *P. chamae* has 3 zoeal stages, while *P. ostreum* has 4.

Megalopa distinguishing characteristics: Antennae with 5 segments. Total of 3 pairs of pleopods on abdominal somites 2–4.

Description:

Zoea, general: Carapace lacks dorsal and lateral spines. Rostral spine greatly reduced. Short lateral spines on abdominal somites 2–3. Telson somewhat squarish and composed of three tooth-like lobes. With 3 pairs of setae between lateral and median lobes of telson.

Zoea I: No additional features.

Zoea II: With 6 swimming setae on each maxilliped. With 3 setae on dorsal surface of abdominal somite 1.

Zoea III: With 8 swimming setae on each maxilliped. With 5 setae on dorsal surface of abdominal somite 1.

Megalopa: Carapace without spines. Antennae with 5 segments. Pleopods on abdominal somites 2–4.

Pinnotheres ostreum Say, 1817 (figs. 38a, b)

References: Hyman (1924), Sandoz & Hopkins (1947).

Synopsis: With 4 zoeal stages. Larval lifespan 25 days at 23°C. Carapace lengths for zoeae I–III and megalopa = 0.42, 0.57, 0.60 and 0.60 mm, respectively.

Zoea distinguishing characteristics: Carapace without dorsal or lateral spines. Rostral spine greatly reduced.

Distinguishing between zoeae of *Pinnotheres chamae* and *P. ostreum*: see *P. chamae*.

Megalopa distinguishing characteristics: Antennae with 5 segments. A total of 4 pairs of pleopods on abdominal somites 2–5.

Description:

Zoea, general: Carapace without dorsal and lateral spines. Rostral spine greatly reduced. Antennae very short, and not visible in some stages. Short lateral spines on abdominal somites 2 and 3. Telson somewhat squarish and composed of three tooth-like lobes. Three pairs of setae between lateral and median lobes of telson.

Zoea I: No additional features.

Zoea II: Pereiopods visible beneath carapace. With 6 swimming setae on each maxilliped.

Zoea III: With 8 swimming setae on each maxilliped. Abdominal somites 2–5 with pleopod buds.

Zoea IV: Undescribed.

Megalopa: Carapace without spines, but with 4–9 setae on lateral margins. Antenna with 5 segments. Pleopods on abdominal somites 2–5.

Pinnotheres maculatus Say, 1818 (figs. 39a, b)

References: Costlow & Bookhout (1966c).

Synopsis: With 5 zoeal stages.

Zoea distinguishing characteristics: Dorsal spine the longest carapace spine.

Megalopa distinguishing characteristics: Carapace with dorsal spine and 2 anteriorly projecting spines. With 3 stiff hairs on dactyl of pereiopod 5.

Description:

Zoea, general: Carapace with dorsal, rostral, and lateral spines. Dorsal spine the longest carapace spine. Lateral spines deflected downward (especially in later zoeal stages). Antennae much shorter than rostral spine. Small lateral spines on abdominal somites 2 and 3. Telson bifurcate without dorsal or lateral spines and with 3 + 3 inner spines. Lumpy process at the base of the telson. Chromatophore pattern: (1) a pair at the base of rostral spine, (2) a pair medial and dorsal to eyes, (3) base of dorsal spine, (4) posterior surface of dorsal spine, (5) base of lateral spines extending dorsally, (6) dorsal to the gut, (7) mandibles and labrum, (8) basipodite of maxilliped 1, (9) abdominal somite 1 dorsal to the gut, (10) postero-ventral surface of abdominal somites 2–5.

Zoea I: No additional features.

Zoea II: With 6 swimming setae on each maxilliped. Lateral spines extend below margin of carapace.

Zoea III: With 8 swimming setae on each maxilliped.

Zoea IV: With 9 swimming setae on maxilliped 1 and 10 on maxilliped 2. With 6 abdominal somites. Small pleopod buds on abdominal somites 2–5.

Zoea V: Pereiopods visible below carapace. Endopodite of antenna longer than protopodite. Pleopod buds biramous.

Megalopa: Postero-dorsal spine and two anteriorly projecting horns on carapace. Rostrum terminates as two rounded lateral projections. Pleopods on abdominal somites 2–6. With 3 stiff hairs on dactyl of pereiopod 5. Chromatophore pattern: (1) a pair median to eyestalks, (2) dorsal surface of eyestalks, (3) posterior to eyestalks, (4) postero-lateral surface of carapace, (5) base of dorsal spine, (6) abdominal somite 1 dorsal to the gut extending into carapace, (7) postero-lateral borders of abdominal somites 2–6.

Pinnixa chaetopterana Stimpson, 1860 (figs. 40, 44)

References: Hyman (1924), Sandifer (1972).

Synopsis: With 5 zoeal stages.

Zoea distinguishing characteristics: Telson with spear-shaped median tooth. Innermost pair of spines on telson short, about half the length of the middle pair of spines (or less).

Distinguishing among zoeae of *Pinnixa* spp.: Although the zoeae of the four described species of *Pinnixa* are superficially similar, they can be distinguished by telson structure (fig. 44). *P. chaetopterana* and *P. cristata* both have a median tooth on the telson, while *P. cylindrica* and *P. sayana* lack this tooth. In *P. chaetopterana*, the innermost pair of spines on the telson (the ones closest to the median tooth and farthest from the tips of the prongs) are short, and are half the length (or less) of the middle pair of spines. In *P. cristata*, the innermost pair of spines is approximately equal in length to the middle pair of spines. There is also a pronounced V-shaped notch between the inner spines on the telson and the median tooth in *P. cristata*; this notch is greatly reduced or absent in *P. chaetopterana*. In *P. cylindrica,* the width between the tips of the prongs is ≥ the width of the telson. In *P. sayana,* the width between the tips of the prongs is < the width of the telson.

Megalopa distinguishing characteristics: Undescribed.

Description:

Zoea, general: Larva caltrop-shaped. Carapace with dorsal, rostral, and lateral spines. Antennae much smaller than rostral spine. Small lateral knobs on abdominal somites 2 and 3. Abdominal somite 5 has lateral knobs that appear as large "wing-like" projections that overlap the telson. Telson bifurcate with spear-shaped median tooth and small lateral spines at the base of each prong. Telson with 3 + 3 inner spines. Innermost pair of spines on the telson (closest to the median tooth and farthest from the tips of the prongs) are short, and are about half the length of the middle pair of spines.

Zoea I: No additional features.

Zoea II: With 6 swimming setae on each maxilliped.

Zoea III: Pereiopods visible beneath carapace. With 8 swimming setae on each maxilliped.

Zoea IV: With 10 swimming setae on each maxilliped. Pleopod buds on abdominal somites 2–5.

Zoea V: Pleopod buds very large and segmented.

Pinnixa cristata Rathbun, 1900 (figs. 41, 44)

References: Dowds (1980).
Synopsis: Only zoea I described.
Zoea distinguishing characteristics: Telson with spear-shaped median tooth. Innermost pair of spines approximately equal in length to middle pair of spines.
Megalopa distinguishing characteristics: Undescribed.
Distinguishing among zoeae of *Pinnixa* spp.: See *P. chaetopterana*.
Description:

Zoea, general: Larva caltrop-shaped. Carapace with dorsal, rostral, and lateral spines. Antennae much smaller than rostral spine. Small lateral knobs on abdominal somites 2 and 3. Abdominal somite 5 has lateral knobs that appear as large "wing-like" projections that overlap the telson. Telson bifurcate with a spear-shaped median tooth and small lateral spines at the base of each prong. Telson with 3 + 3 inner spines. Innermost pair of spines are approximately equal in length to the middle pair of spines. Pronounced V-shaped notch between the inner spines and the median tooth on the telson.

Zoea I: No additional features.

Pinnixa cylindrica (Say, 1818) (figs. 42, 44)

References: Hyman (1924), Sandifer (1972).
Synopsis: Only zoea I described.
Zoea distinguishing characteristics: Telson without median tooth. Distance between the tips of the prongs ≥ telson width.
Distinguishing among zoeae of *Pinnixa* spp.: See *P. chaetopterana*.
Megalopa distinguishing characteristics: Undescribed.

Description:

Zoea, general: Larvae caltrop-shaped. Carapace with dorsal, rostral, and lateral spines. Lateral spines curve ventrally. Antennae much smaller than rostral spine. Small lateral knobs on abdominal somites 2 and 3. Abdominal somite 5 has large lateral knobs that appear as "wing-like" projections that overlap telson. Telson bifurcate with the distance between the tips of the prongs $\geq$ width of telson. Telson with 3 + 3 inner spines.

Zoea I: No additional features.

Pinnixa sayana Stimpson, 1860 (figs. 43, 44)

References: Hyman (1924), Sandifer (1972).

Synopsis: With 5 zoeal stages.

Zoea distinguishing characteristics: Telson without median tooth. Distance between the tips of the prongs < telson width.

Distinguishing between zoeae of *Pinnixa* spp.: See *P. chaetopterana*.

Megalopa distinguishing characteristics: Undescribed.

Description:

Zoea, general: Larvae caltrop-shaped. Carapace with dorsal, rostral, and lateral spines. Small lateral knobs on abdominal somites 2 and 3. Abdominal somite 5 has lateral knobs that appear as large "wing-like" projections that overlap telson. Telson bifurcate with the distance between the tips of the prongs < width of telson. Telson with 3 + 3 inner spines.

Zoea I: No additional features.

Zoea II: With 6 swimming setae on each maxilliped.

Zoea III: Pereiopods visible beneath the carapace. With 8 swimming setae on each maxilliped.

Zoea IV: With 10 swimming setae on each maxilliped. Pleopod buds on abdominal somites 2–5.

Zoea V: Pleopod buds very large and segmented.

Sesarma cinereum (Bosc, 1802) (figs. 45a, b)

References: Costlow & Bookhout (1960).

Synopsis: With 4 zoeal stages.

Zoea distinguishing characteristics: Carapace with dorsal and rostral spines, but without lateral spines. Antennae approximately equal in length to rostral spine.

Distinguishing between zoeae of *S. reticulatum* and *S. cinereum*: Zoeae I of *Sesarma reticulatum* and *S. cinereum* may be distinguished by the setation on the endopodite of maxilliped 1 (see Costlow & Bookhout (1962b) for an in-depth discussion of this technique). Additionally, *S. reticulatum* has an endopodite knob on the base of the antennae, while *S. cinereum* does not. Beyond the zoea I, however, distinguishing between species becomes confounded by the different number of zoeal stages in the two species. *S. cinereum* has 4 zoeal stages, while *S. reticulatum* has 3. Investigators should pay close attention to the number of free abdominal somites and the number of swimming setae on the maxillipeds, as it will often be possible to differentiate between species using these characteristics (again, see Costlow & Bookhout, 1962, for a more detailed description of how to differentiate between these species).

Megalopa distinguishing characteristics: Rostrum reduced. With 3 long hairs on dactyl of pereiopod 5. Antennae with 9 segments.

Description:

Zoea, general: Carapace with dorsal and rostral spines, but without lateral spines. Antennae approximately equal in length to rostral spine. Small knob on abdominal somite 2. Spine on the border of abdominal somite 3. Lateral spines on posterior margins of abdominal somites 2–5 that overlap following somite. Telson bifurcate without dorsal or lateral spines. Chromatophore pattern: (1) median to each eye, (2) a single frontal chromatophore (unicorn-like), (3) ventral to the heart, (4) mandibles and labrum, (5) basipodite of maxilliped 1, (6) abdominal somite 1 dorsal to the gut, (7) postero-ventral borders of abdominal somites 2–5 (and somite 6 when added).

Zoea I: Telson with 3 + 3 inner spines.

Zoea II: With 6 swimming setae on each maxilliped.

Zoea III: Pereiopods visible beneath carapace. With 8 swimming setae on each maxilliped. With 6 abdominal somites. Pleopod buds on abdominal somites 2–5.

Zoea IV: Pereiopods well developed. With 9 swimming setae on maxilliped 1 and 10 on maxilliped 2. Pleopods well developed. Telson with 4 + 4 inner spines.

Megalopa: Center of rostrum depressed. Rostral spine absent. Antennae with 9 segments. Three stiff spines on dactyls of pereiopod 5. Small dorsal spines on abdominal somites 2–6. Telson without lateral spines. Chromatophore pattern: (1) two on dorsal surface of rostrum, (2) anterior and posterior peripheries of eyestalks, (3) lateral borders of carapace, (4) posterior edge of carapace dorsal to the gut, (5) mandibles and labrum, (6) dorso-lateral surfaces of all abdominal somites.

Sesarma reticulatum (Say, 1817) (figs. 46a, b)

References: Costlow & Bookhout (1962b).

Synopsis: With 3 zoeal stages.

Zoea distinguishing characteristics: Carapace with dorsal and rostral spines, but without lateral spines. Antennae approximately equal in length to rostral spine.

Distinguishing between zoeae of *S. reticulatum* and *S. cinereum*: See *S. cinereum*.

Megalopa distinguishing characteristics: Rostrum long, pointed, and directed anteriorly.

Description:

Zoea, general: Carapace with dorsal and rostral spines, but lacking lateral spines. Antennae approximately equal in length to rostral spine. Small knob on abdominal somite 2. Spine on border of abdominal somite 3. Telson bifurcate without dorsal or lateral spines and with 3 + 3 inner spines. Chromatophore pattern: (1) median to each eye, (2) a single frontal chromatophore (unicorn-like), (3) posterior to dorsal spine, (4) lateral border of carapace, (5) mandibles and labrum, (6) base of rostral spine, (7) base of antennules, (8) abdominal somite 1 dorsal to the gut, (9) postero-ventral borders of abdominal somites 2–5 (and 6 when added).

Zoea I: No additional features.

Zoea II: With 6 swimming setae on each maxilliped. With 6 abdominal somites. Postero-lateral spines on abdominal somites 2–6 that overlap the following somite. Pleopod buds present.

Zoea III: Pereiopods visible beneath carapace. With 8 swimming setae on each maxilliped. Pleopod buds better developed.

Megalopa: Carapace with long rostral spine. Antennae composed of 7 segments. With 3 long setae on dactyl of pereiopod 5. With 6 abdominal somites. Telson with one pair of large lateral spines and 3 pairs of inner setae. Chromatophore pattern: (1) 2 chromatophores median to the eyes on dorsal surface of rostrum, (2) antero-lateral margin of carapace, (3) several on postero-lateral margin of carapace, (4) a single chromatophore on posterior edge of carapace dorsal to the gut, (5) a pair on dorsal-lateral margin of each abdominal somite, (6) mandible and labrum.

Ocypode quadrata (Fabricius, 1787) (figs. 47a, b)

References: Diaz & Costlow (1972).
Synopsis: With 5 zoeal stages. Larval lifespan 34 days to megalopa at 25°C.
Zoea distinguishing characteristics: Carapace caltrop-shaped with rounded protuberance on the central region.
Megalopa distinguishing characteristics: Pereiopod 4 carried dorsal to carapace.
Description:
   Zoea, general: Zoea caltrop-shaped. Carapace with dorsal, rostral and lateral spines. Dorsal spine nearly straight and approximately the same length as rostral spine. Antennae with spinules. Distinctive protuberance on central frontal region of carapace. Postero-lateral margins of abdominal somites 2–5 end in blunt points that overlap next somites. Telson bifurcate with no lateral spines.
   Zoea I: Telson with 3 + 3 inner spines.
   Zoea II: With 6 swimming setae on each maxilliped. Telson with 4 + 4 inner spines.
   Zoea III: Carapace depressions and protrusions appear in pairs. With 8 swimming setae on each maxilliped. With 6 abdominal somites. Postero-lateral spines of somites 4–6 appear as long spines. Telson with 5 + 5 inner spines.
   Zoea IV: With 6 plumose setae on abdominal somite 1. With 10 swimming setae on each maxilliped. Pleopod buds on abdominal somites 2–6. Telson with 6 + 6 inner spines.
   Zoea V: Pereiopods well developed. With 11 plumose setae on abdominal somite 1. Pleopods on abdominal somites 4–6. Telson with 7 + 7 inner spines.
   Megalopa: Carapace oval, rounded dorsally, and with slight depressions and prominences. Rostrum deflected downward. Postero-lateral regions of carapace bear depressions in which the last pair of legs fit. Pereiopod 4 lies in lateral depressions over the other pereiopods with the dactyls hooking over the eyestalk. With 6 abdominal somites. Pleopods on abdominal somites 2–6. Terminal end of telson somewhat pointed with 15 plumose setae along central margin.

Uca spp. (figs. 48a, b)

References: Hyman (1920), Sandifer (1972).
Synopsis: With 5 zoeal stages. Total length of zoea I = 1.00 mm.
Zoea distinguishing characteristics: Carapace with dorsal and rostral spines, but without lateral spines. Rostral spine longer than antennae.
Distinguishing among zoeae of *Uca* species: According to Hyman (1920), it is impossible to differentiate among zoeae of different *Uca* species. Citing an

unpublished manuscript by Kurata, Sandifer (1972) states that it might be possible to distinguish *Uca* spp. using chromatophore patterns, but provides no details about these techniques.

Megalopa distinguishing characteristics: Antennae with 11 segments.

Description:

Zoea, general: Larva fairly small. Carapace with dorsal and rostral spines, but without lateral spines. Antennae about half the length of the rostrum. Lateral spines on abdominal somites 2 and 3. Postero-lateral spines on abdominal somites 2–5. Telson bifurcate without lateral or dorsal spines.

Zoea I: Telson with 3 + 3 inner spines.

Zoea II: With 6 swimming setae on each maxilliped.

Zoea III: With 8 swimming setae on each maxilliped. With 6 abdominal somites. Telson with 4 + 4 inner spines.

Zoea IV: With 9 swimming setae on each maxilliped. Pleopod buds on abdominal somites 2–6.

Zoea V: With 10 swimming setae on each maxilliped. Pleopod buds better developed.

Megalopa: Rostrum ends in two blunt, knob-like lateral processes. Carapace without spines or processes. Antennae with 11 segments. Pereiopod 5 carried on dorsal surface of carapace. Postero-lateral spines on abdominal somite 5. Telson rounded.

ACKNOWLEDGEMENTS

Funding for this project was provided by an NSF pre-doctoral fellowship to S. G. Bullard and by a Publication and Research grant from Wake Forest University. I thank H. Eure, M. Hay, J. C. von Vaupel Klein, N. Lindquist, F. Schwartz, J. Stachowicz, E. Sotka, E. Cruz-Rivera, G. Cetrulo, L. Manning, and J. Grabowski for their support during this project. Special thanks to J. D. Costlow, Jr. and P. A. Sandifer for their assistance and willingness to allow me to republish figures from their original manuscripts. Comments by G. Fisher and an anonymous reviewer greatly improved this manuscript.

APPENDIX I

Differentiating among the zoeal stages of spider crabs (Majidae)

Gore et al. (1982) provide an excellent discussion about how to differentiate among the zoeal and megalopal stages of the subfamily Mithracinae (excluding the zoeal stages of *Stenocionops*, which were described by Laughlin et al., 1984). While much of their discussion is pertinent to this key, some of the zoeae discussed come from crab species not found in North Carolina waters. In addition, the zoeal stages of *Libinia dubia* and *L. emarginata* are sufficiently similar to those of the Mithracinae to make identification difficult. Table I draws upon and expands the discussion of Gore et al. (1982), however, investigators are referred to that work for additional information.

Coloration and chromatophore pattern of the Majidae
(paraphrased from Gore et al., 1982)

Microphrys zoeae are predominantly lime or olive green with scattered melanophores, whereas *Mithrax* zoeae are generally transparent, highlighted with yellow, gold, red orange, or rose coloration, with scattered melanophores. In *Microphrys bicornutus*, a prominent black chromatophore occurs distally on the first, and proximally on the second, of each of the lime green maxilliped basipodites. In *Mithrax forceps* (A. Milne-Edwards, 1875) the basipodites are clear or red orange, and the melanophore is more diffuse and located medioventrally. A similar pattern of coloration is seen in *M. pleuracanthus* Stimpson, 1871 but the zoeae have "a great deal of yellow" on the carapace and abdomen (Goy et al., 1981). The chromatophores on the carapace are red in other *Mithrax* zoeae. The general color of *Macrocoeloma camptocerum* (Stimpson, 1871) is a light yellow green with several melanophores on the basipodite of maxilliped 1 and the gastric region.

TABLE I

Morphological characteristics of North Carolina Pisinae and Mithracinae (derived from Gore et al., 1982). Ratio of the dorsal carapace spine indicates the comparative size of the spine relative to carapace height. Rostral carapace spine length indicates the approximate size of the rostral spine relative to the antennule. Maxillule and maxilliped endopodite setae refer to the number and grouping of setae on these appendages

Species	Carapace spines		Antennule: # of aesthetascs + # of setae	Antenna: Exopodite : protopodite	Maxillule: Endopodite setae	Maxilla: # Setae on endopodite	Maxilliped setae	
	Dorsal	Rostral					Endopodite	# Setae on basipodite
Libinia dubia	0.7, slightly curved	1/2	4 + 2	exo = proto	1, 1 + 4	5	3, 2, 1, 2, 4 + 1	3
Libinia emarginata	0.6, slightly hooked	1/2	4 + 1	exo = proto	1, 1 + 4	5	3, 2, 1, 2, 4 + 1	3
Microphrys bicornutus	0.6, slightly curved	>1/2	4 + 1	exo ≥ proto	1, 2 + 4	5	3, 2, 1, 2, 4 + 1	3
Mithrax forceps	0.7, moderately curved	1/2	4 + 1	exo ≥ proto	1, 2 + 4	5	3, 2, 1, 2, 4 + 1	3
Mithrax pleuracanthus	0.6, curved	1/2	3 + 1	exo = proto	1, 2 + 4	5	3, 2, 1, 2, 4 + 1	3
Mithrax spinosissimus	0.4, slightly hooked	1/3	5 + 0	exo < proto	2 terminal setae	1	0, 1, 1, 2, 3 + 1	0
Macrocoeloma camptocerum	0.8, oblique	<1/2	3 + 1	exo = proto	1, 1 + 4	5	3, 2, 1, 2, 4 + 1	3
Stenocionops furcatus coelatus	0.4, strongly curved	<1/2	3 + 0	exo < proto	1, 2 + 4	5	2, 2, 1, 2, 5 + 0	3

APPENDIX II

Tables II, III and IV describe characteristics (setation and dorsal spine lengths) that can be used to differentiate between zoeal and megalopal stages of *Callinectes sapidus* and *C. similis*.

TABLE II

Larval setation of *Callinectes sapidus* (from Bookhout & Costlow, 1977); pld, plumodenticulate; pl, plumose

Larval stage	Maxilla: Coxal endite, distal lobe		Scaphognathite	Maxilliped 1	Maxilliped 2 exopodites
	Terminal	Subterminal			
Zoea I	1 pld	1 pld, 2 pl	5	4	4
Zoea II	1 pld	1 pld, 2 pl	8	6	6
Zoea III	1 pld	1 pld, 2 pl	13	8	8
Zoea IV	1 pld	1 pld, 2 pl	15	9	10
Zoea V	1 pld	1 pld, 2 pl	20-21	10	11
Zoea VI	1 pld	2 pld, 2 pl	25	11	13
Zoea VII	1 pld	2 pld, 2 pl	26-29	12	14
Zoea VIII	1 pld	2 pld, 2 pl	33-36	13	15
Megalopa	1 pld	4 pl			

TABLE III

Larval setation of *Callinectes similis* (from Bookhout & Costlow, 1977); pld, plumodenticulate; pl, plumose

Larval stage	Maxilla: Coxal endite, distal lobe		Scaphognathite	Maxilliped 1	Maxilliped 2 exopodites
	Terminal	Subterminal			
Zoea I	1 short	1 pld, 2 pl	5	4	4
Zoea II	1 short	1 pld, 2 pl	8	6	6
Zoea III	1 pld	1 pld, 2 pl	13	8	8
Zoea IV	1 pld	1 pld, 2 pl	17-20	9	9
Zoea V	1 pld	1 pld, 2 pl	23	11	12
Zoea VI	1 pld	2 pld, 2 pl	26-28	12	14
Zoea VII	1 pld	2 pld, 2 pl	31-33	14	15
Zoea VIII	1 pld	2 pld, 2 pl	40	15	17
Megalopa	1 pld	5 pl			

TABLE IV

Length of the dorsal spines (mm) of *Callinectes sapidus* and *C. similis* zoeae (derived from Bookhout & Costlow, 1977)

Larval stage	C. sapidus	C. similis
Zoea I	0.289	0.341
Zoea II	0.359	0.375
Zoea III	0.405	0.437
Zoea IV	0.475	0.615
Zoea V	0.575	0.778
Zoea VI	0.652	0.995
Zoea VII	0.828	0.993
Zoea VIII	0.940	1.205

58

REFERENCES

ANDRYSZAK, B. L. & R. H. GORE, 1981. The complete larval development in the laboratory of *Micropanope sculptipes* (Crustacea, Decapoda, Xanthidae) with a comparison of larval characters in western Atlantic xanthid genera. Fish. Bull., U.S., **79**: 487-506.

BOOKHOUT, C. G. & J. D. COSTLOW, JR., 1974. Larval development of *Portunus spinicarpus* reared in the laboratory. Bull. Mar. Sci., **24** (1): 20-51.

—— & ——, 1977. Larval development of *Callinectes similis* reared in the laboratory. Bull. Mar. Sci., **27** (4): 704-728.

—— & ——, 1979. Larval development of *Pilumnus dasypodus* and *Pilumnus sayi* reared in the laboratory (Decapoda Brachyura, Xanthidae). Crustaceana, (Suppl.) **5**: 1-16.

CHAMBERLAIN, N. A., 1961. Studies on the larval development of *Neopanope texana sayi* (Smith) and other crabs of the family Xanthidae (Brachyura). Chesapeake Bay Inst., Techn. Rep., **22**: 1-35.

——, 1962. Ecological studies of the larval development of *Rhithropanopeus harrisii* (Xanthidae, Brachyura). Chesapeake Bay Inst., Techn. Rep., **28**: 1-47.

COSTLOW, J. D., JR. & C. G. BOOKHOUT, 1959. The larval development of *Callinectes sapidus* Rathbun reared in the laboratory. Biol. Bull.,Woods Hole, **116** (3): 373-396.

—— & ——, 1960. The complete larval development of *Sesarma cinereum* (Bosc) reared in the laboratory. Biol. Bull., Woods Hole, **118** (2): 203-214.

—— & ——, 1961a. The larval development of *Eurypanopeus depressus* (Smith) under laboratory conditions. Crustaceana, **2** (1): 6-15.

—— & ——, 1961b. The larval stages of *Panopeus herbstii* Milne-Edwards reared in the laboratory. Journ. Elisha Mitchell Sci. Soc., **77** (1): 33-42.

—— & ——, 1962a. The larval development of *Hepatus epheliticus* (L.) under laboratory conditions. Journ. Elisha Mitchell Sci. Soc., **78** (2): 113-125.

—— & ——, 1962b. The larval development of *Sesarma reticulatum* Say reared in the laboratory. Crustaceana, **4** (4): 281-294.

—— & ——, 1966a. Larval development of the crab *Hexapanopeus angustifrons*. Chesapeake Sci., **7** (3): 148-156.

—— & ——, 1966b. The larval development of *Ovalipes ocellatus* (Herbst) under laboratory conditions. Journ. Elisha Mitchell Sci. Soc., **82** (2): 160-171.

—— & ——, 1966c. Larval stages of the crab, *Pinnotheres maculatus*, under laboratory conditions. Chesapeake Sci., **7** (3): 157-163.

—— & ——, 1968. Larval development of the crab, *Leptodius agassizii* A. Milne Edwards in the laboratory (Brachyura, Xanthidae). Crustaceana, (Suppl.) **2**: 203-213.

DIAZ, H. & J. D. COSTLOW, JR., 1972. Larval development of *Ocypode quadrata* (Brachyura: Crustacea) under laboratory conditions. Mar. Biol., Berlin, **15** (2): 120-131.

DOWDS, R. E., 1980. The crab genus *Pinnixa* in a North Carolina estuary: identification of larvae, reproduction, and recruitment: 1-189. (Ph.D. Thesis, University of North Carolina, Chapel Hill).

GORE, R. H., 1968. The larval development of the commensal crab *Polyonyx gibbesi* Haig, 1956 (Crustacea: Decapoda). Biol. Bull., Woods Hole, **135** (1): 111-129.

——, 1971. The complete larval development of *Porcellana sigsbeiana* (Crustacea: Decapoda) under laboratory conditions. Mar. Biol., Berlin, **11**: 344-355.

——, 1973. Studies on the decapod Crustacea from the Indian River of Florida. II. *Megalobrachium soriatum* (Say, 1818): the larval development under laboratory culture (Crustacea, Decapoda, Anomura). Bull. Mar. Sci., **23** (4): 837-856.

GORE, R. H., L. E. SCOTTO & W. T. YANG, 1982. *Microphrys bicornutus* (Latreille, 1825): the complete larval development under laboratory conditions with notes on other Mithracinae larvae (Decapoda: Brachyura: Majidae). J. Crust. Biol., **2**: 514-534.

GOY, J. W., C. G. BOOKHOUT & J. D. COSTLOW, JR., 1981. Larval development of the spider crab *Mithrax pleuracanthus* Stimpson reared in the laboratory (Decapoda: Brachyura: Majidae). J. Crust. Biol., **1**: 51-62.

HOOD, M. R., 1962. Studies of the larval development of *Rhithropanopeus harrisii* (Gould) of the family Xanthidae (Brachyura). Gulf. Res. Rep., **1** (3): 122-130.

HYMAN, O. W., 1920. The development of *Gelasimus* after hatching. J. Morphol., **33** (2): 485-501.

— —, 1924. Studies of the larval crabs of the family Pinnotheridae. Proc. U.S. Natl. Museum, **64** (7): 1-7.

INGLE, R. W., 1992. Larval stages of northeastern Atlantic crabs: 1-363. (Chapman & Hall, London).

JOHNS, D. M. & W. H. LANG, 1977. Larval development of the spider crab *Libinia emarginata* (Majidae). Fish. Bull., U.S., **75** (4): 831-841.

KIRCHER, A. B., 1970. The zoeal stages and glaucothoë of *Hypoconcha arcuata* Stimpson (Decapoda: Dromiidae) reared in the laboratory. Bull. Mar. Sci., **20** (3): 769-792.

LAUGHLIN, R., P. RODRÍGUEZ & E. WEIL, 1984. The zoeal stages of the decorator crab *Stenocionops furcatus coelatus* (A. Milne Edwards, 1878) (Decapoda, Oxyrhyncha, Majidae) reared in the laboratory. Crustaceana, **46** (2): 202-208.

NEGREIROS-FRANSOZO, M. L., A. FRANSOZO & N. J. HEBLING, 1989. Larval development of *Persephona mediterranea* (Herbst, 1794) (Brachyura, Leucosiidae) under laboratory conditions. Crustaceana, **57** (2): 177- 193.

PAULA, J., 1996. A key and bibliography for the identification of zoeal stages of brachyuran crabs (Crustacea, Decapoda, Brachyura) from the Atlantic coast of Europe. J. Plankt. Res., **18** (1): 17-27.

POHLE, G. & M. TELFORD, 1981. The larval development of *Dissodactylus crinitichelis* Moreira, 1901 (Brachyura: Pinnotheridae) in laboratory culture. Bull. Mar. Sci., **31**: 753-773.

PORTER, H. J., 1960. Zoeal stages of the stone crab, *Menippe mercenaria* Say. Chesapeake Sci., **1** (3-4): 168-177.

REES, G. H., 1959. Larval development of the sand crab *Emerita talpoida* (Say) in the laboratory. Biol. Bull., Woods Hole, **117** (2): 356-370.

ROBERTS, M. H., JR., 1968. Larval development of the decapod *Euceramus praelongus* in laboratory culture. Chesapeake Sci., **9** (2): 121-130.

— —, 1975. Larval development of *Pinnotheres chamae* reared in the laboratory. Chesapeake Sci., **16** (4): 242-252.

SANDIFER, P. A., 1972. Morphology and ecology of Chesapeake Bay decapod crustacean larvae: 1-532. (Ph.D. Thesis, University of Virginia, Charlottesville).

— —, 1974. Larval stages of the crab, *Pilumnus dasypodus* Kingsley (Crustacea, Brachyura, Xanthidae), obtained from the laboratory. Bull. Mar. Sci., **24** (2): 378-391.

SANDIFER, P. A. & W. A. VAN ENGEL, 1971. Larval development of the spider crab *Libinia dubia* H. Milne Edwards (Brachyura, Majidae, Pisinae), reared in the laboratory. Chesapeake Sci., **12** (1): 18-25.

— — & — —, 1972a. Larval stages of the spider crab *Anasimus latus* Rathbun, 1894 (Brachyura, Majidae, Inachinae) obtained in the laboratory. Crustaceana, **23** (2): 141-151.

— — & — —, 1972b. *Lepidopa* larvae (Crustacea, Decapoda, Albuneidae) from Virginia Plankton. Journ. Elisha Mitchell Sci. Soc., **88** (4): 220-225.

SANDOZ, M. & S. HOPKINS, 1947. Early life history of the oyster crab, *Pinnotheres ostreum* (Say). Biol. Bull., Woods Hole, **93** (3): 250-258.

SASTRY, A. N., 1977a. The larval development of the Jonah crab, *Cancer borealis* Stimpson, 1859, under laboratory conditions (Decapoda, Brachyura). Crustaceana, **32** (3): 290-303.

— —, 1977b. The larval development of the rock crab, *Cancer irroratus* Say, 1817, under laboratory conditions (Decapoda, Brachyura). Crustaceana, **32** (2): 155-168.

WILLIAMS, A. B., 1984. Shrimps, lobsters, and crabs of the Atlantic coast of the eastern United States, Maine to Florida: 1-550. (Smithsonian Institution Press, Washington, D.C.).

FIGURES
4–48b

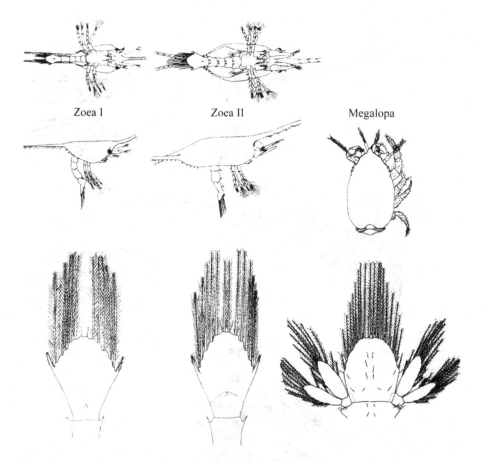

Zoea I   Zoea II   Megalopa

Fig. 4. *Euceramus praelongus*. Top row, ventral views of zoeae. Middle row, side view of zoeae, dorsal view of megalopa. Bottom row, zoeal and megalopal telsons. [Reprinted with permission from Roberts, 1968.]

Zoea I

Zoea II

Megalopa

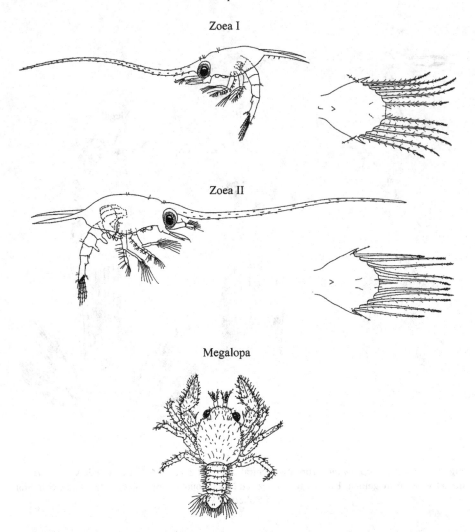

Fig. 5. *Megalobrachium soriatum*. Top row, side view and telson of zoea I. Middle row, side view and telson of zoea II. Bottom row, dorsal view of megalopa. [Reprinted with permission from Gore, 1973.]

Zoea I

Zoea II

Megalopa

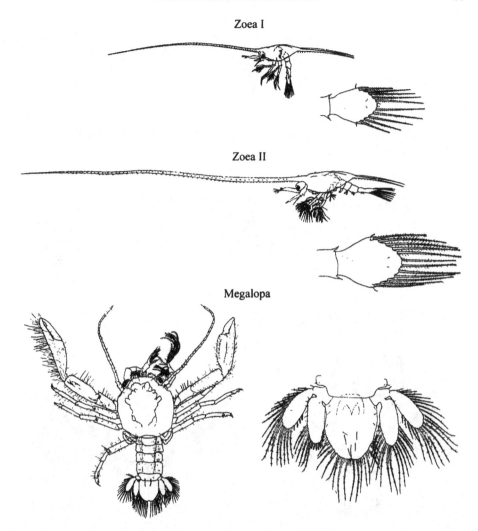

Fig. 6. *Polyonyx gibbesi*. Top row, side view and telson of zoea I. Middle row, side view and telson of zoea II. Bottom row, dorsal view and telson of megalopa. [Reprinted with permission from Gore, 1968.]

Zoea I

Zoea II

Megalopa

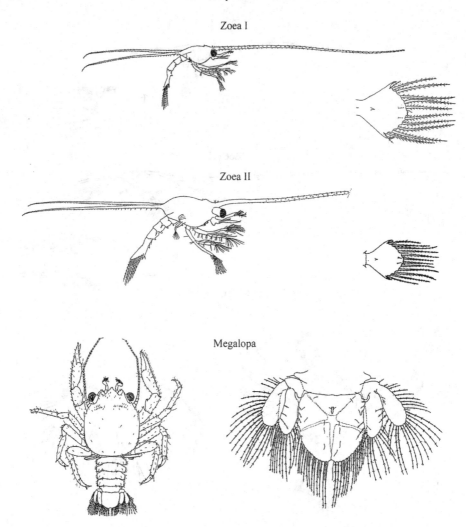

Fig. 7. *Porcellana sigsbeiana*. Top row, side view and telson of zoea I. Middle row, side view and telson of zoea II. Bottom row, dorsal view and telson of megalopa. [Copyright permission sought from Gore, 1971; figs. 2, 3, 6, copyright held by Springer-Verlag, Berlin and Heidelberg.]

Zoea I

Zoea II

Zoea III

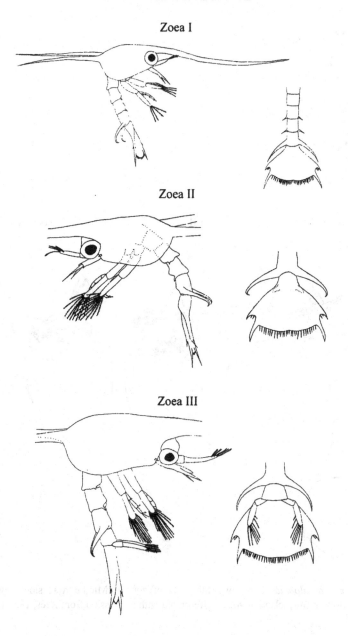

Fig. 8. *Lepidopa* sp. Top row, side view and telson of zoea I. Middle row, side view and telson of zoea II. Bottom row, side view and telson of zoea III. [Reprinted with permission from Sandifer & Van Engel, 1972b.]

Zoea I

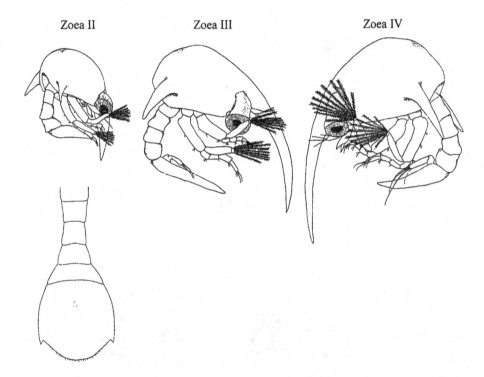

Fig. 9a. *Emerita talpoida*. Top row, side view of zoea I. Middle row, side view of zoeae. Bottom row, telson of zoea I. [Reprinted with permission from Rees, 1959.]

Zoea V   Zoea VI   Megalopa

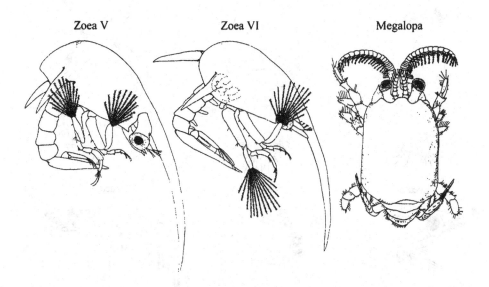

Fig. 9b. *Emerita talpoida*. Side view of zoeae, dorsal view of megalopa. [Reprinted with permission from Rees, 1959.]

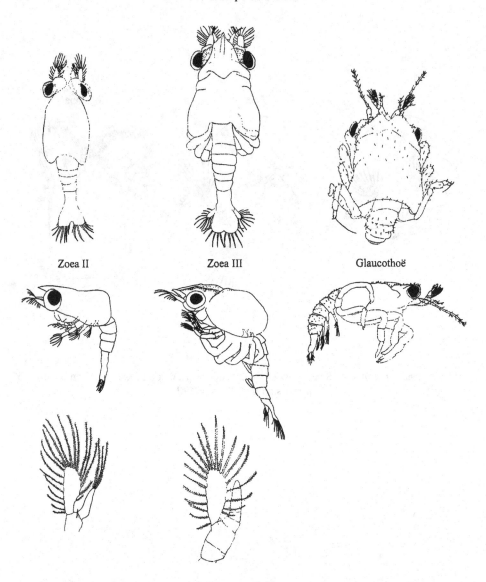

Fig. 10. *Hypoconcha arcuata*. Top row, dorsal view of zoeae and megalopa. Middle row, side view of zoeae and megalopa. Bottom row, zoeal antennae. [Reprinted with permission from Kircher, 1970.]

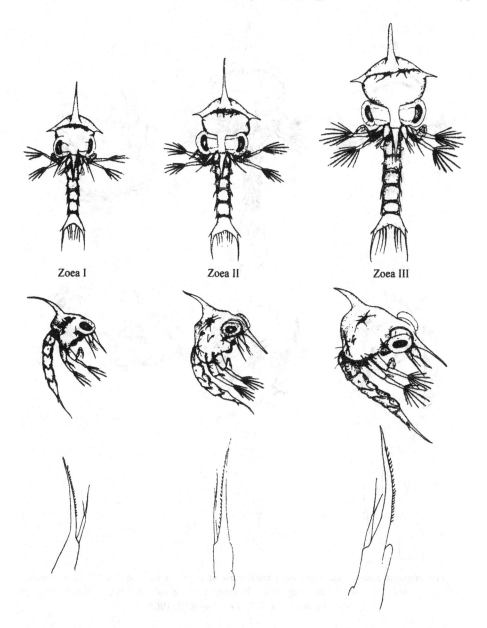

Zoea I Zoea II Zoea III

Fig. 11a. *Hepatus epheliticus*. Top row, ventral view of zoeae. Middle row, side view of zoeae. Bottom row, zoeal antennae. [Reprinted with permission from Costlow & Bookhout, 1962a.]

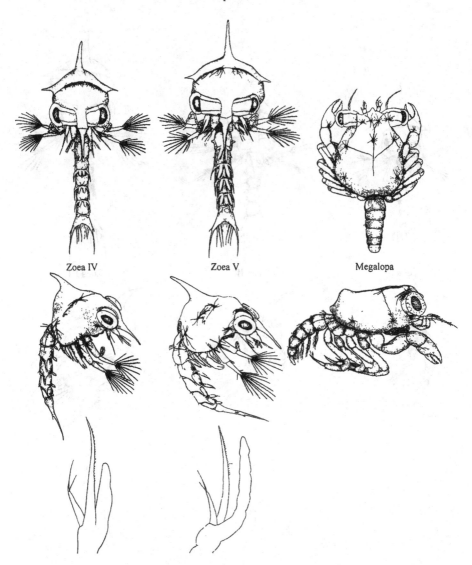

Zoea IV   Zoea V   Megalopa

Fig. 11b. *Hepatus epheliticus.* Top row, ventral view of zoeae, dorsal view of megalopa. Middle row, side view of zoeae and megalopa. Bottom row, zoeal antennae. [Reprinted with permission from Costlow & Bookhout, 1962a.]

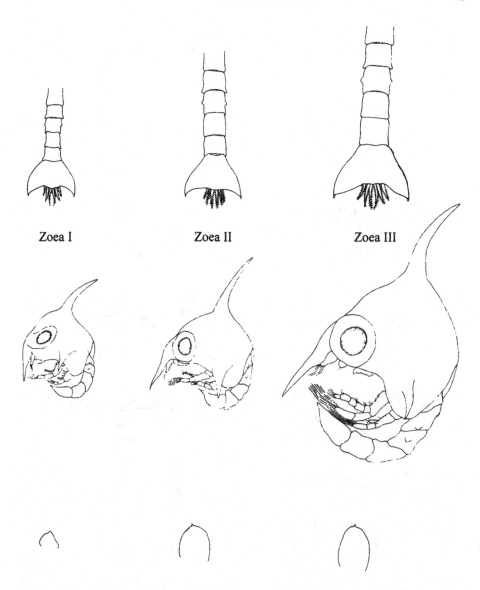

Fig. 12a. *Persephona mediterranea.* Top row, zoeal telsons. Middle row, side view of zoeae. Bottom row, zoeal antennae. [Reprinted with permission from Negreiros-Fransozo et al., 1989.]

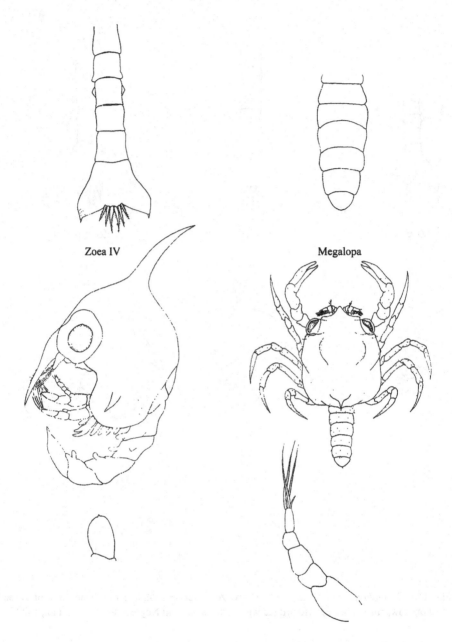

Zoea IV   Megalopa

Fig. 12b. *Persephona mediterranea.* Top row, zoeal and megalopal telsons. Middle row, side view of zoea, dorsal view of megalopa. Bottom row, zoeal and megalopal antennae. [Reprinted with permission from Negreiros-Fransozo et al., 1989.]

Zoea I   Zoea II

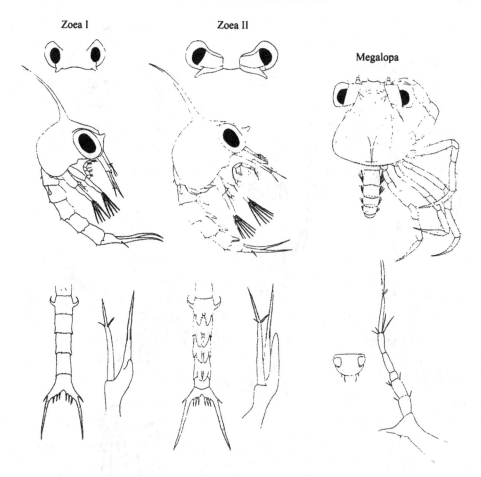

Megalopa

Fig. 13. *Anasimus latus*. Top row, the eyes of zoeae showing reduced rostral spines. Middle row, side view of zoeae, dorsal view of megalopa. Bottom row, zoeal and megalopal telsons and antennae. [Reprinted with permission from Sandifer & Van Engel, 1972a.]

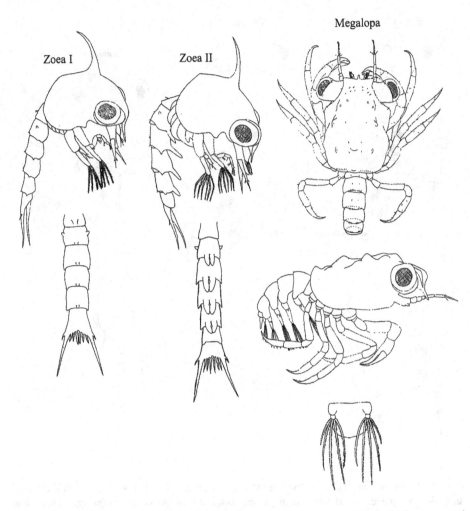

Fig. 14. *Libinia dubia*. Top row, side view of zoeae, dorsal view of megalopa. Middle row, zoeal telsons, side view of megalopa. Bottom row, megalopal telson. [Reprinted with permission from Sandifer & Van Engel, 1971.]

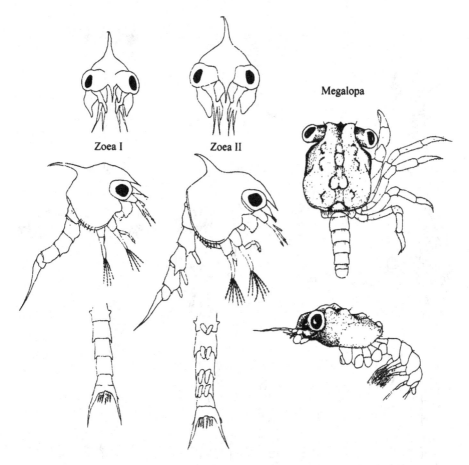

Megalopa

Zoea I   Zoea II

Fig. 15. *Libinia emarginata*. Top row, frontal view of zoeal carapace. Middle row, side view of zoeae, dorsal view of megalopa. Bottom row, zoeal telsons, side view of megalopa. [Reprinted with permission from Johns & Lang, 1977.]

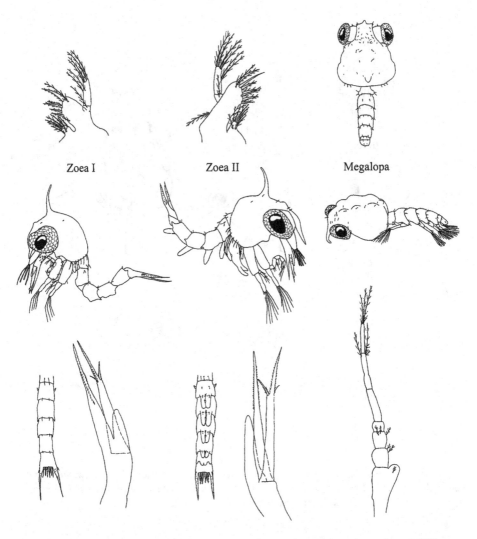

Zoea I   Zoea II   Megalopa

Fig. 16. *Microphrys bicornutus*. Top row, zoeal maxillulae, dorsal view of megalopa. Middle row, side view of zoeae and megalopa. Bottom row, zoeal telsons and antennae, megalopal antenna. [Reprinted with permission from Gore et al., 1982.]

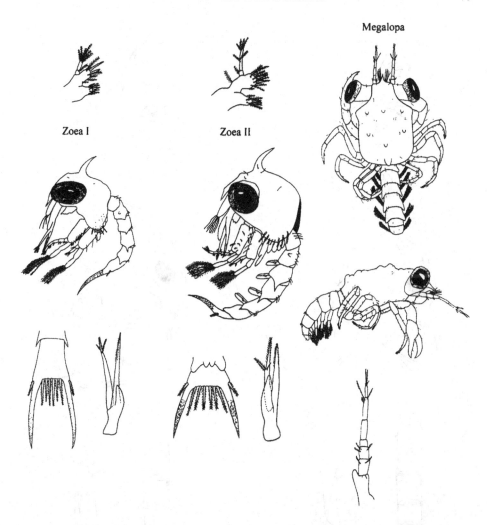

Fig. 17. *Mithrax pleuracanthus*. Top row, zoeal maxillulae, dorsal view of megalopa. Middle row, side view of zoeae and megalopa. Bottom row, zoeal telsons and antennae, megalopal antenna. [Reprinted with permission from Goy et al., 1981.]

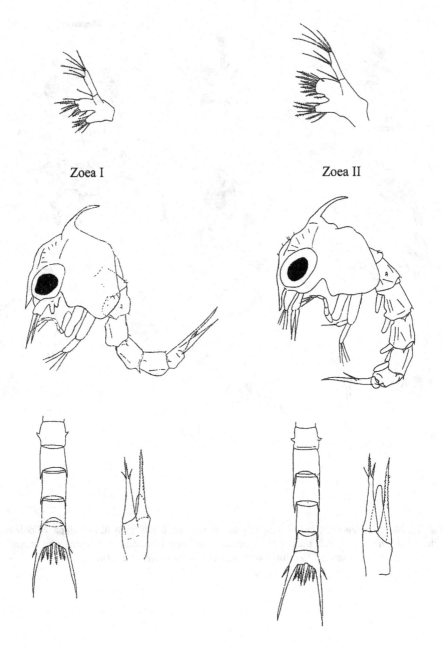

Zoea I   Zoea II

Fig. 18. *Stenocionops furcatus coelatus*. Top row, zoeal maxillulae. Middle row, side view of zoeae. Bottom row, zoeal telsons and antennae. [Reprinted with permission from Laughlin et al., 1984.]

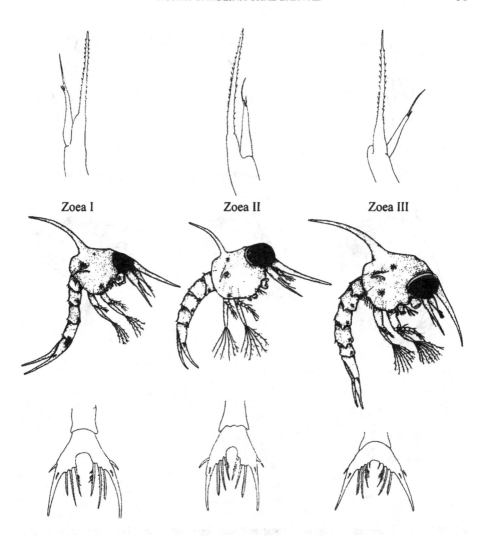

Zoea I   Zoea II   Zoea III

Fig. 19a. *Cancer irroratus.* Top row, zoeal antennae. Middle row, side view of zoeae. Bottom row, zoeal telsons. [Reprinted with permission from Sastry, 1977b.]

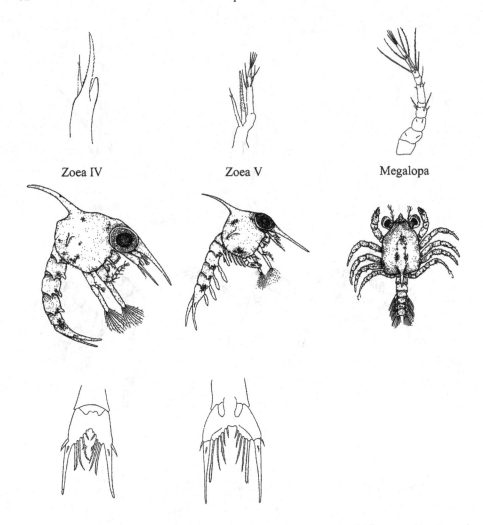

Zoea IV   Zoea V   Megalopa

Fig. 19b. *Cancer irroratus*. Top row, zoeal and megalopal antennae. Middle row, side view of zoeae, dorsal view of megalopa. Bottom row, zoeal telsons. [Reprinted with permission from Sastry, 1977b.]

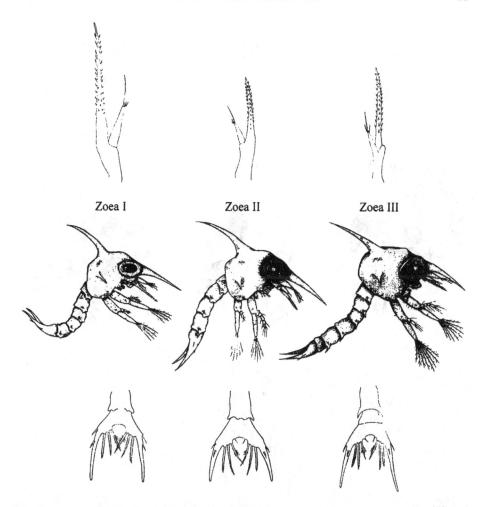

Zoea I   Zoea II   Zoea III

Fig. 20a. *Cancer borealis*. Top row, zoeal antennae. Middle row, side view of zoeae. Bottom row, zoeal telsons. [Reprinted with permission from Sastry, 1977a.]

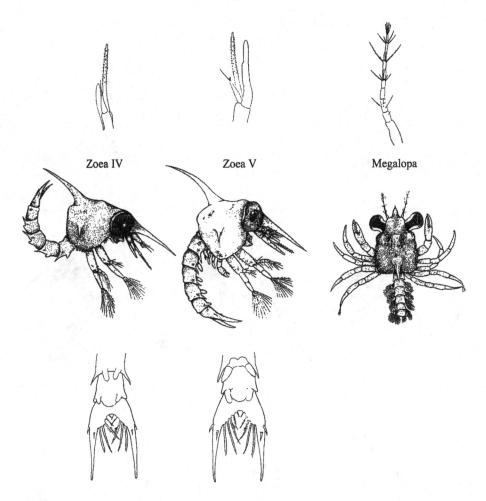

Fig. 20b. *Cancer borealis*. Top row, zoeal and megalopal antennae. Middle row, side view of zoeae, dorsal view of megalopa. Bottom row, zoeal telsons. [Reprinted with permission from Sastry, 1977a.]

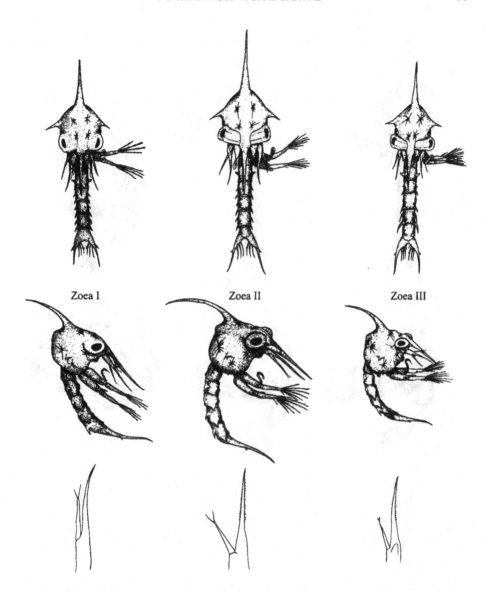

Fig. 21a. *Ovalipes ocellatus*. Top row, ventral view of zoeae. Middle row, side view of zoeae. Bottom row, zoeal antennae. [Reprinted with permission from Costlow & Bookhout, 1966b.]

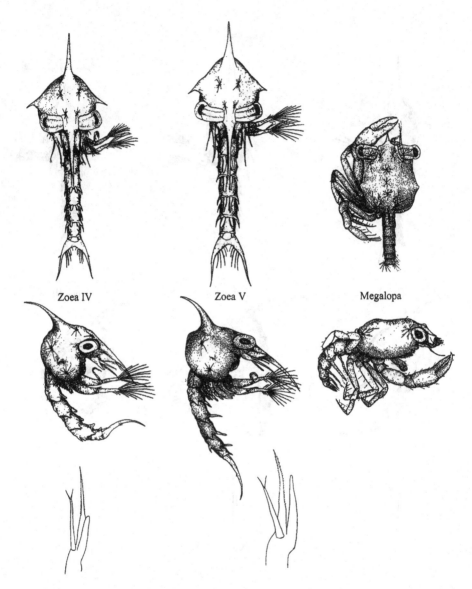

Zoea IV   Zoea V   Megalopa

Fig. 21b. *Ovalipes ocellatus*. Top row, ventral view of zoeae, dorsal view of megalopa. Middle row, side view of zoeae and megalopa. Bottom row, zoeal antennae. [Reprinted with permission from Costlow & Bookhout, 1966b.]

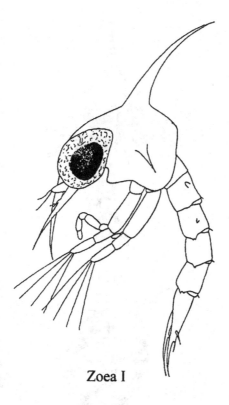

Zoea I

Fig. 22. *Arenaeus cribrarius*. Top row, side view of zoea. Bottom row, zoeal telson. [Reprinted with permission from Sandifer, 1972.]

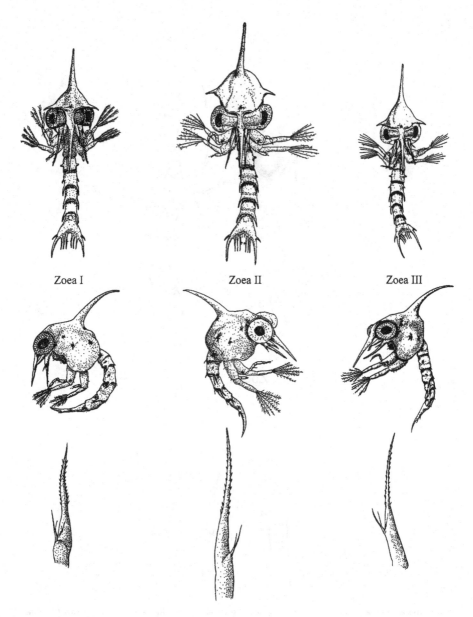

Fig. 23a. *Callinectes sapidus.* Top row, ventral view of zoeae. Middle row, side view of zoeae. Bottom row, zoeal antennae. [Reprinted with permission from Costlow & Bookhout, 1959.]

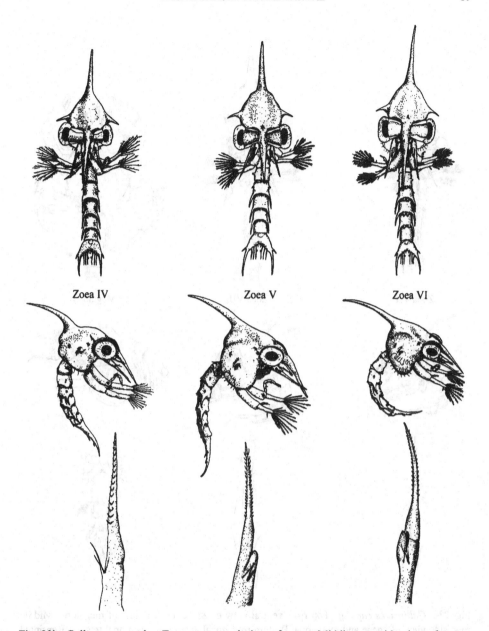

Zoea IV   Zoea V   Zoea VI

Fig. 23b. *Callinectes sapidus*. Top row, ventral view of zoeae. Middle row, side view of zoeae. Bottom row, zoeal antennae. [Reprinted with permission from Costlow & Bookhout, 1959.]

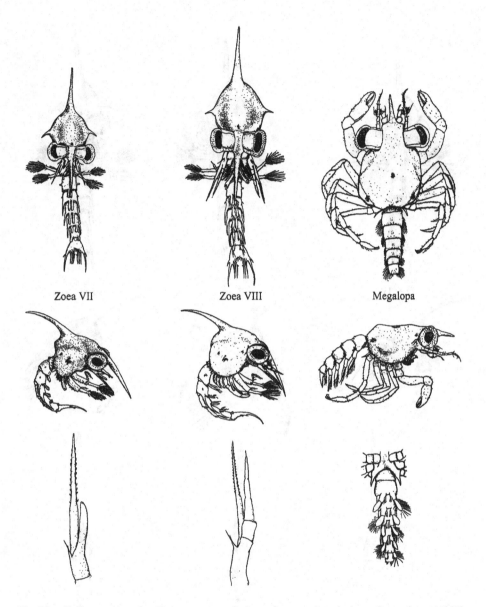

Zoea VII   Zoea VIII   Megalopa

Fig. 23c. *Callinectes sapidus*. Top row, ventral view of zoeae, dorsal view of megalopa. Middle row, side view of zoeae and megalopa. Bottom row, zoeal antennae, megalopal telson. [Reprinted with permission from Costlow & Bookhout, 1959.]

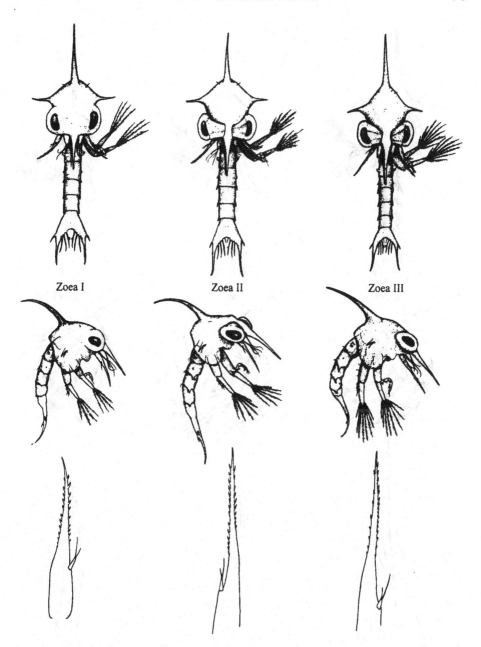

Zoea I   Zoea II   Zoea III

Fig. 24a. *Callinectes similis*. Top row, ventral view of zoeae. Middle row, side view of zoeae. Bottom row, zoeal antennae. [Reprinted with permission from Bookhout & Costlow, 1977.]

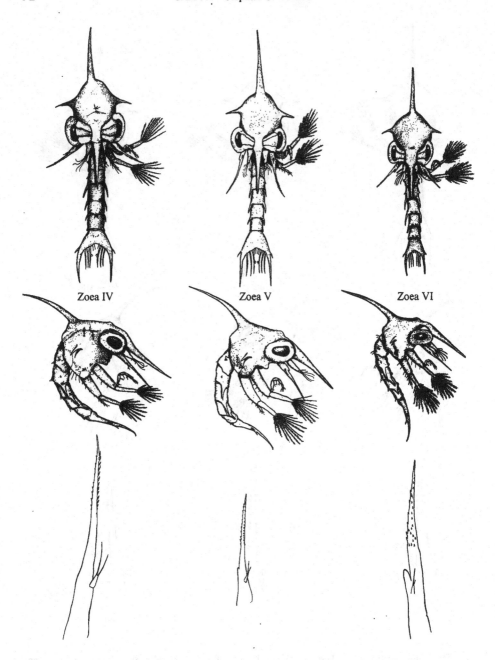

Fig. 24b. *Callinectes similis*. Top row, ventral view of zoeae. Middle row, side view of zoeae. Bottom row, zoeal antennae. [Reprinted with permission from Bookhout & Costlow, 1977.]

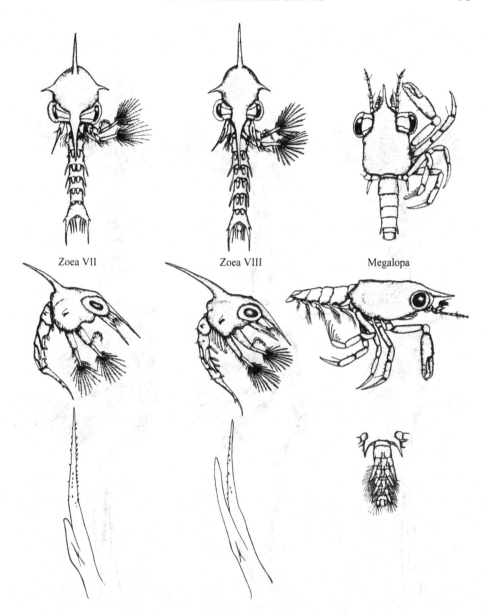

Zoea VII   Zoea VIII   Megalopa

Fig. 24c. *Callinectes similis*. Top row, ventral view of zoeae, dorsal view of megalopa. Middle row, side view of zoeae and megalopa. Bottom row, zoeal antennae, megalopal telson. [Reprinted with permission from Bookhout & Costlow, 1977.]

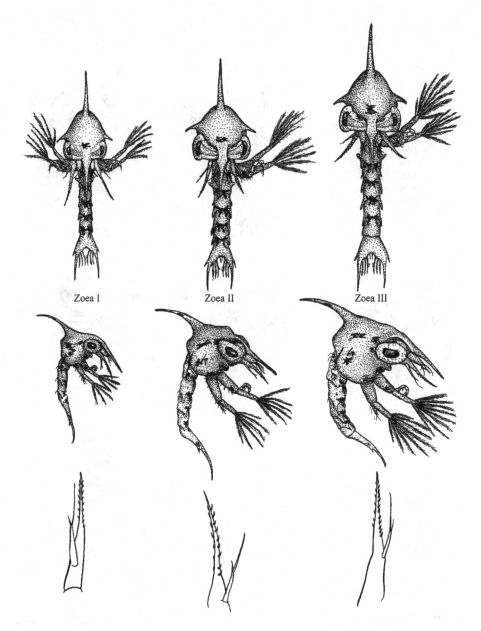

Fig. 25a. *Portunus spinicarpus*. Top row, ventral view of zoeae. Middle row, side view of zoeae. Bottom row, zoeal antennae. [Reprinted with permission from Bookhout & Costlow, 1974.]

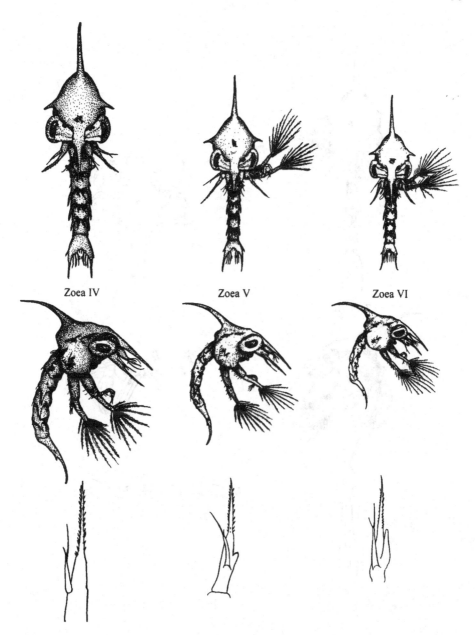

Zoea IV   Zoea V   Zoea VI

Fig. 25b. *Portunus spinicarpus*. Top row, ventral view of zoeae. Middle row, side view of zoeae. Bottom row, zoeal antennae. [Reprinted with permission from Bookhout & Costlow, 1974.]

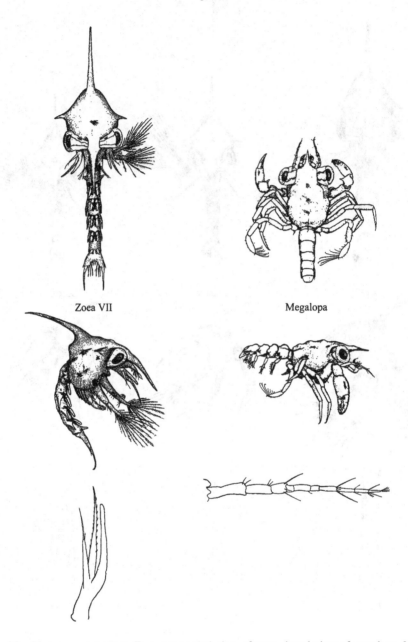

Fig. 25c. *Portunus spinicarpus*. Top row, ventral view of zoea, dorsal view of megalopa. Middle row, side view of zoea and megalopa. Bottom row, zoeal and megalopal antennae. [Reprinted with permission from Bookhout & Costlow, 1974.]

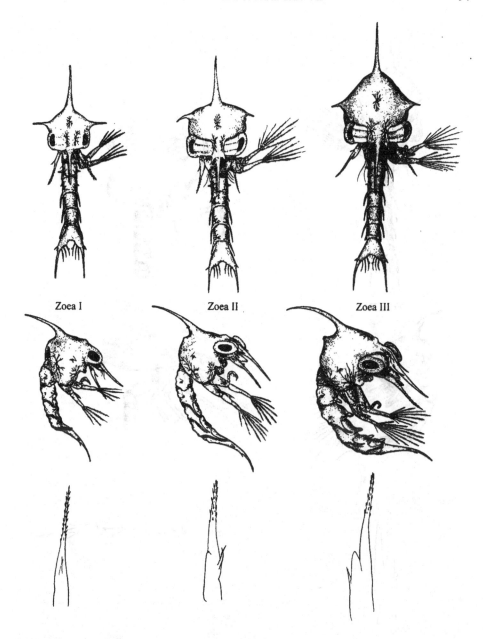

Zoea I   Zoea II   Zoea III

Fig. 26a. *Pseudomedaeus agassizii*. Top row, ventral view of zoeae. Middle row, side view of zoeae. Bottom row, zoeal antennae. [Reprinted with permission from Costlow & Bookhout, 1968.]

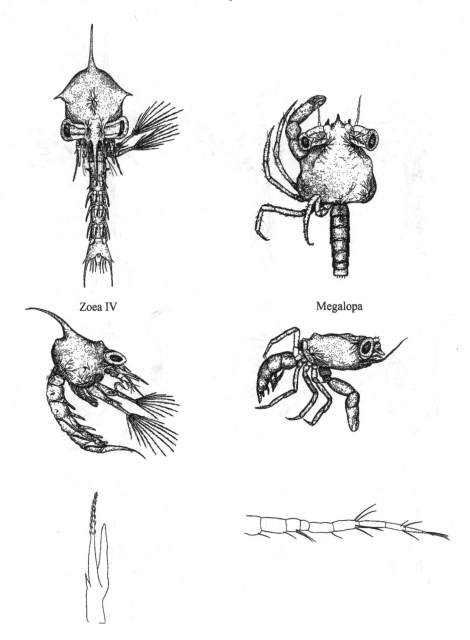

Fig. 26b. *Pseudomedaeus agassizii*. Top row, ventral view of zoea, dorsal view of megalopa. Middle row, side view of zoea and megalopa. Bottom row, zoeal and megalopal antennae. [Reprinted with permission from Costlow & Bookhout, 1968.]

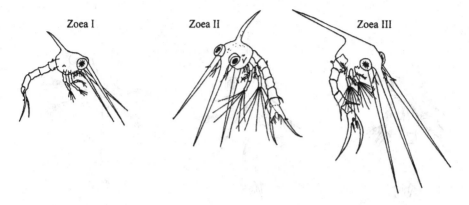

Fig. 27a. *Rhithropanopeus harrisii.* Side view of zoeae. [Copyright permission sought from Hood, 1962.]

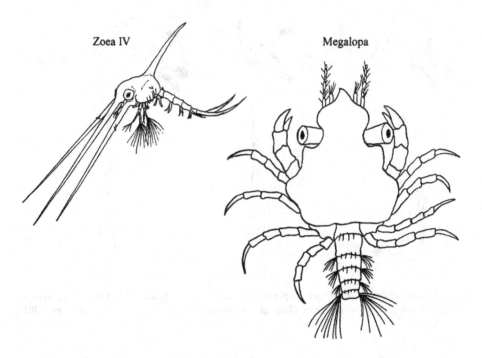

Fig. 27b. *Rhithropanopeus harrisii.* Side view of zoea, dorsal view of megalopa. [Copyright permission sought from Hood, 1962.]

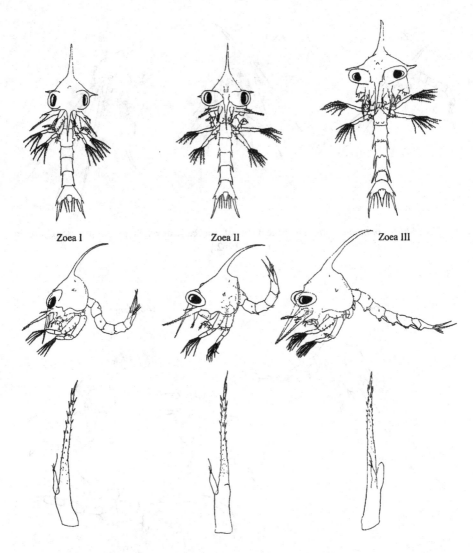

Fig. 28a. *Micropanope sculptipes*. Top row, ventral view of zoeae. Middle row, side view of zoeae. Bottom row, zoeal antennae. [Reprinted with permission from Andryszak & Gore, 1981.]

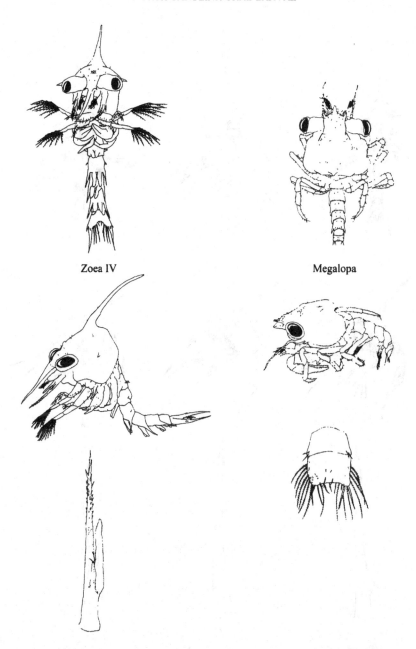

Zoea IV   Megalopa

Fig. 28b. *Micropanope sculptipes*. Top row, ventral view of zoea, dorsal view of megalopa. Middle row, side view of zoea and megalopa. Bottom row, zoeal antenna, megalopal telson. [Reprinted with permission from Andryszak & Gore, 1981.]

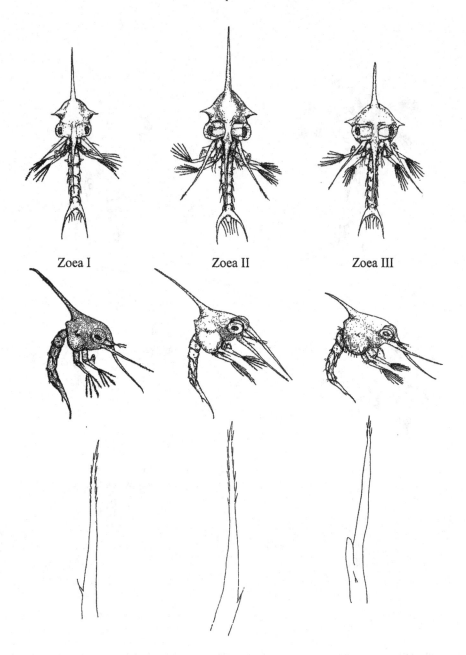

Zoea I   Zoea II   Zoea III

Fig. 29a. *Eurypanopeus depressus*. Top row, ventral view of zoeae. Middle row, side view of zoeae. Bottom row, zoeal antennae. [Reprinted with permission from Costlow & Bookhout, 1961a.]

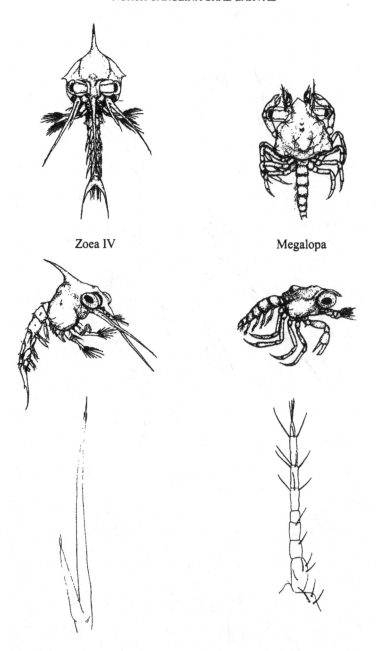

Zoea IV   Megalopa

Fig. 29b. *Eurypanopeus depressus*. Top row, ventral view of zoea, dorsal view of megalopa. Middle row, side view of zoea and megalopa. Bottom row, zoeal and megalopal antennae. [Reprinted with permission from Costlow & Bookhout, 1961a.]

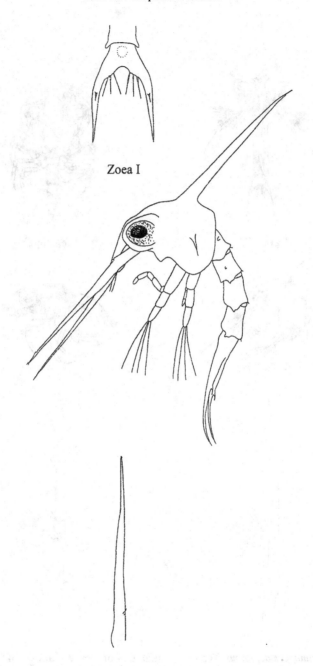

Zoea I

Fig. 30. *Dyspanopeus sayi*. Top row, zoeal telson. Middle row, side view of zoea. Bottom row, zoeal antenna. [Reprinted with permission from Sandifer, 1972.]

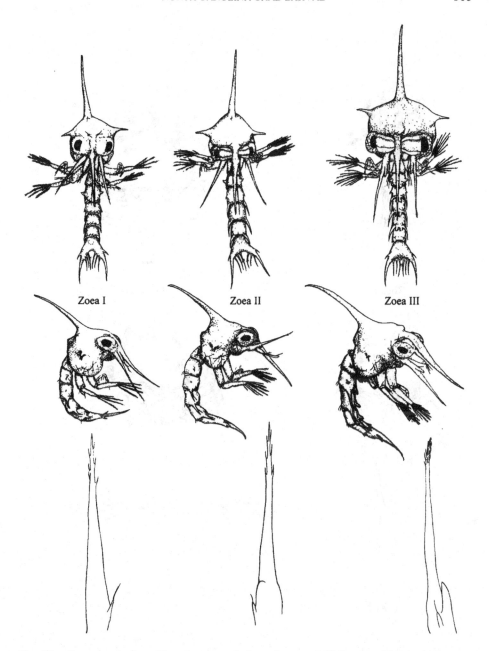

Zoea I Zoea II Zoea III

Fig. 31a. *Panopeus herbstii*. Top row, ventral view of zoeae. Middle row, side view of zoeae. Bottom row, zoeal antennae. [Reprinted with permission from Costlow & Bookhout, 1961b.]

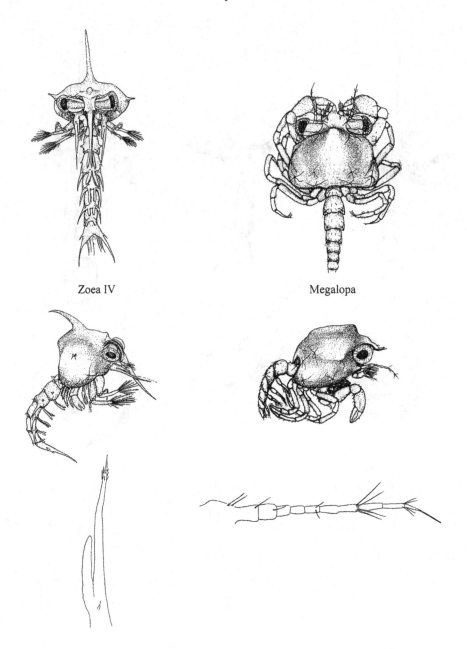

Zoea IV   Megalopa

Fig. 31b. *Panopeus herbstii*. Top row, ventral view of zoea, dorsal view of megalopa. Middle row, side view of zoea and megalopa. Bottom row, zoeal and megalopal antennae. [Reprinted with permission from Costlow & Bookhout, 1961b.]

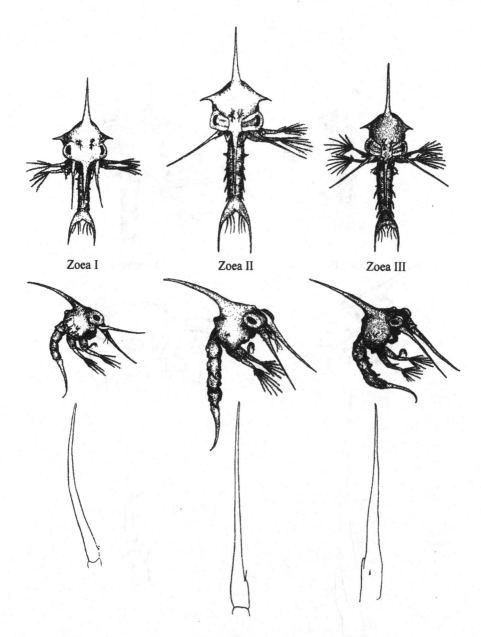

Zoea I   Zoea II   Zoea III

Fig. 32a. *Hexapanopeus angustifrons*. Top row, ventral view of zoeae. Middle row, side view of zoeae. Bottom row, zoeal antennae. [Reprinted with permission from Costlow & Bookhout, 1966a.]

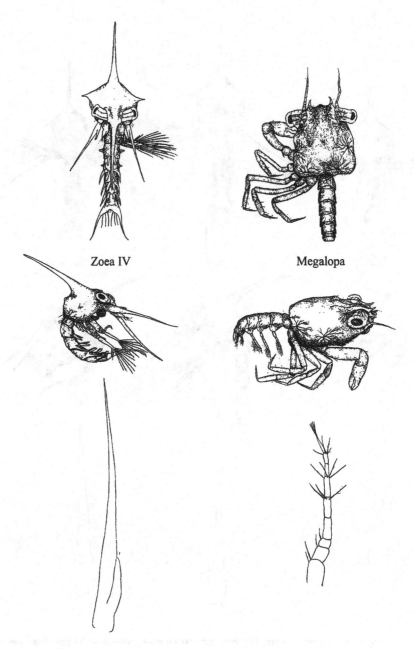

Zoea IV   Megalopa

Fig. 32b. *Hexapanopeus angustifrons*. Top row, ventral view of zoea, dorsal view of megalopa. Middle row, side view of zoea and megalopa. Bottom row, zoeal and megalopal antennae. [Reprinted with permission from Costlow & Bookhout, 1966a.]

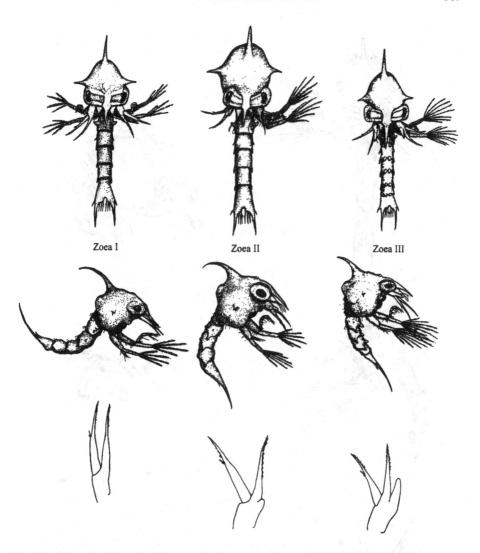

Zoea I   Zoea II   Zoea III

Fig. 33a. *Pilumnus dasypodus.* Top row, ventral view of zoeae. Middle row, side view of zoeae. Bottom row, zoeal antennae. [Reprinted with permission from Bookhout & Costlow, 1979.]

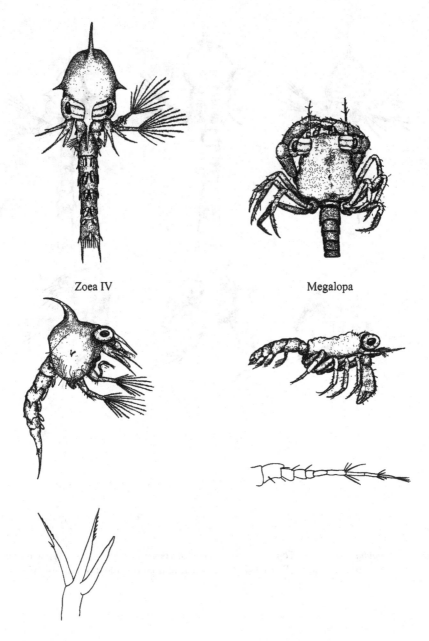

Zoea IV

Megalopa

Fig. 33b. *Pilumnus dasypodus*. Top row, ventral view of zoea, dorsal view of megalopa. Middle row, side view of zoea and megalopa. Bottom row, zoeal and megalopal antennae. [Reprinted with permission from Bookhout & Costlow, 1979.]

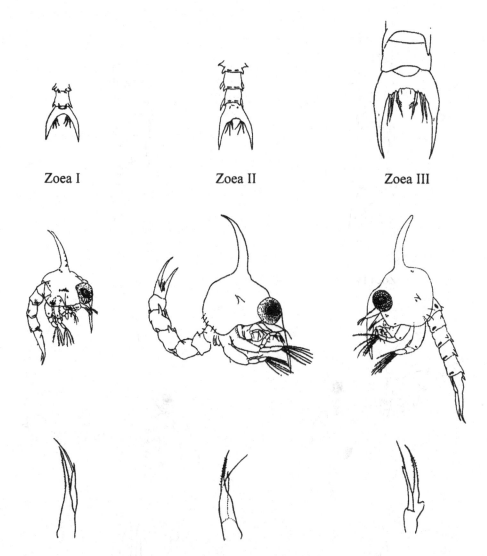

Zoea I   Zoea II   Zoea III

Fig. 34a. *Menippe mercenaria*. Top row, zoeal telsons. Middle row, side view of zoeae. Bottom row, zoeal antennae. [Reprinted with permission from Porter, 1960.]

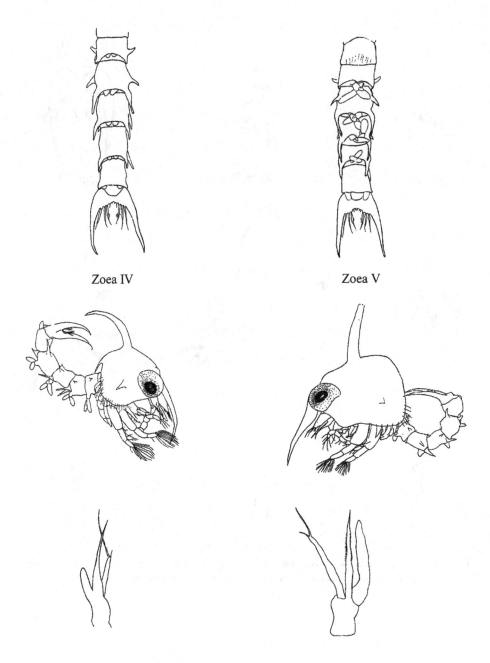

Zoea IV   Zoea V

Fig. 34b. *Menippe mercenaria*. Top row, zoeal telsons. Middle row, side view of zoeae. Bottom row, zoeal antennae. [Reprinted with permission from Porter, 1960.]

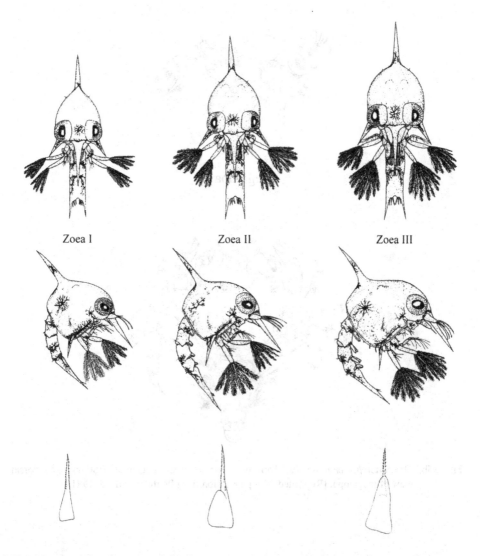

Zoea I   Zoea II   Zoea III

Fig. 35a. *Dissodactylus crinitichelis.* Top row, ventral view of zoeae. Middle row, side view of zoeae. Bottom row, zoeal antennae. [Reprinted with permission from Pohle & Telford, 1981.]

Megalopa

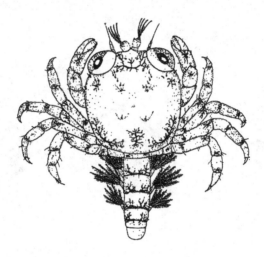

Fig. 35b. *Dissodactylus crinitichelis*. Top row, side view of megalopa. Bottom row, dorsal view of megalopa. [Reprinted with permission from Pohle & Telford, 1981.]

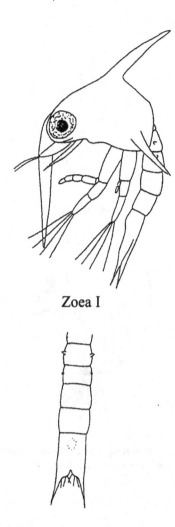

Zoea I

Fig. 36. *Dissodactylus mellitae.* Top row, side view of zoea. Bottom row, zoeal telson. [Reprinted with permission from Sandifer, 1972.]

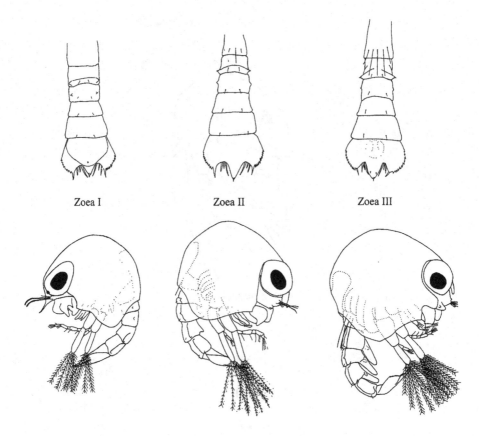

Zoea I   Zoea II   Zoea III

Fig. 37a. *Pinnotheres chamae*. Top row, zoeal telsons. Bottom row, side view of zoeae. [Reprinted with permission from Roberts, 1975.]

Megalopa

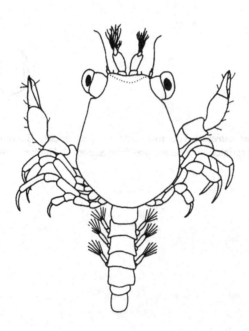

Fig. 37b. *Pinnotheres chamae*. Top row, megalopal antenna. Bottom row, dorsal view of megalopa. [Reprinted with permission from Roberts, 1975.]

Zoea I   Zoea II   Zoea III

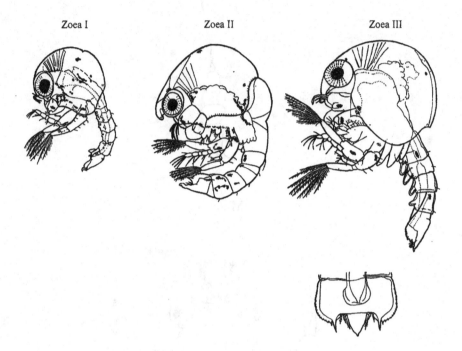

Fig. 38a. *Pinnotheres ostreum*. Top row, side view of zoeae. Bottom row, zoea III telson. [Reprinted with permission from Sandoz & Hopkins, 1947.]

Megalopa

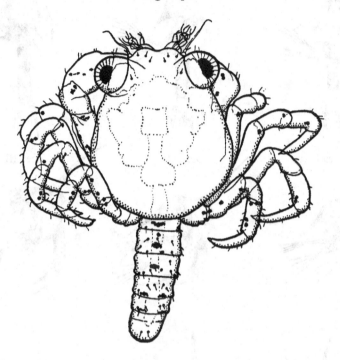

Fig. 38b. *Pinnotheres ostreum.* Dorsal view of megalopa. [Reprinted with permission from Sandoz & Hopkins, 1947.]

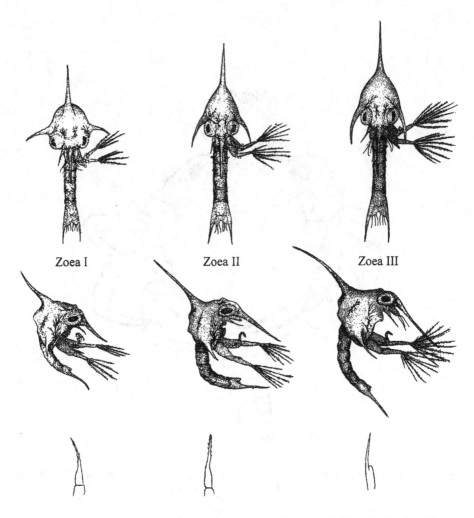

Zoea I Zoea II Zoea III

Fig. 39a. *Pinnotheres maculatus*. Top row, ventral view of zoeae. Middle row, side view of zoeae. Bottom row, zoeal antennae. [Reprinted with permission from Costlow & Bookhout, 1966c.]

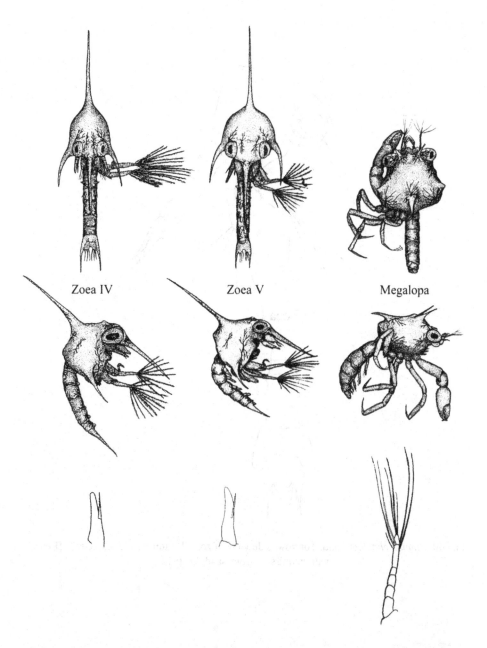

Fig. 39b. *Pinnotheres maculatus*. Top row, ventral view of zoeae, dorsal view of megalopa. Middle row, side view of zoeae and megalopa. Bottom row, zoeal and megalopal antennae. [Reprinted with permission from Costlow & Bookhout, 1966c.]

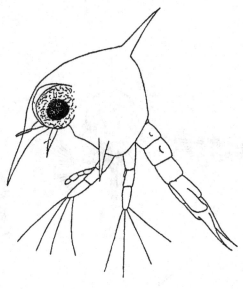

Zoea I

Fig. 40. *Pinnixa chaetopterana*. Top row, side view of zoea. Bottom row, zoeal telson. [Reprinted with permission from Sandifer, 1972.]

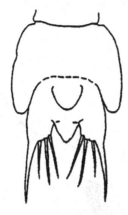

Zoea I

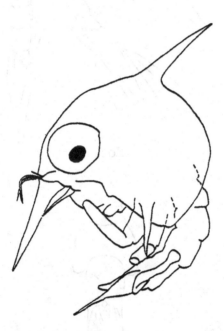

Fig. 41. *Pinnixa cristata*. Top row, zoeal telson. Bottom row, side view of zoea. [Copyright permission sought from Dowds, 1980.]

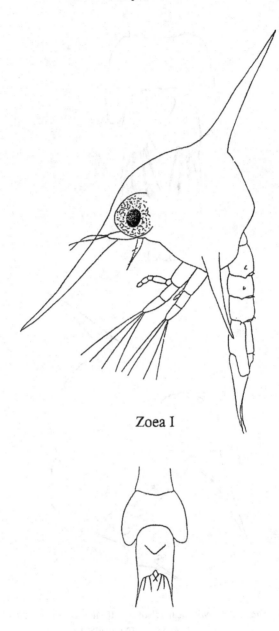

Zoea I

Fig. 42. *Pinnixa cylindrica*. Top row, side view of zoea. Bottom row, zoeal telson. [Reprinted with permission from Sandifer, 1972.]

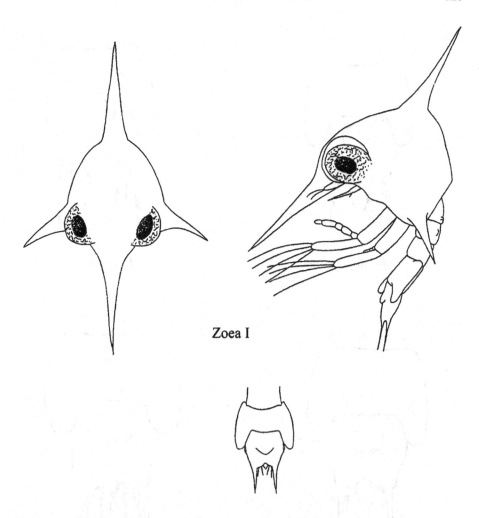

Zoea I

Fig. 43. *Pinnixa sayana*. Top row, frontal view of zoeal carapace, side view of zoea. Bottom row, zoeal telson. [Reprinted with permission from Sandifer, 1972.]

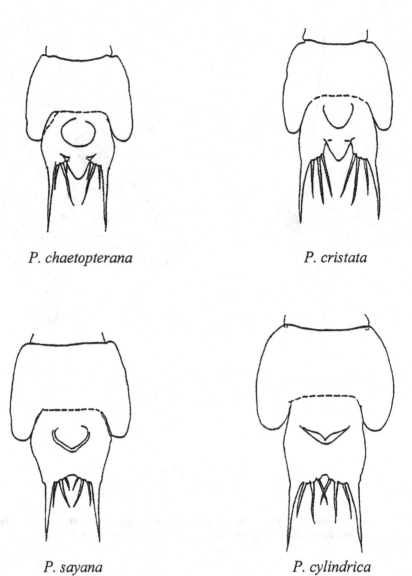

P. chaetopterana

P. cristata

P. sayana

P. cylindrica

Fig. 44. Comparison of the ventral surface of the telsons of *Pinnixa* zoeae. [Copyright permission sought from Dowds, 1980.]

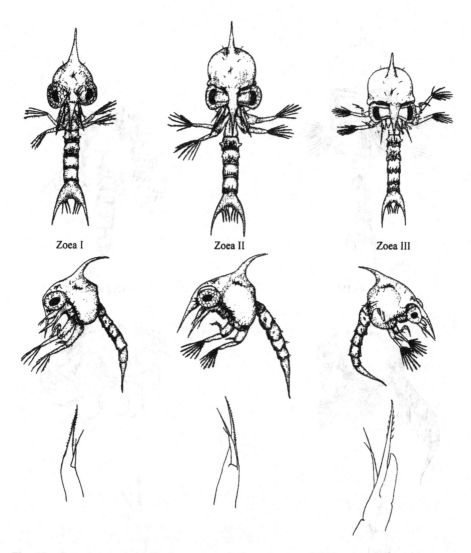

Zoea I   Zoea II   Zoea III

Fig. 45a. *Sesarma cinereum*. Top row, ventral view of zoeae. Middle row, side view of zoeae. Bottom row, zoeal antennae. [Reprinted with permission from Costlow & Bookhout, 1960.]

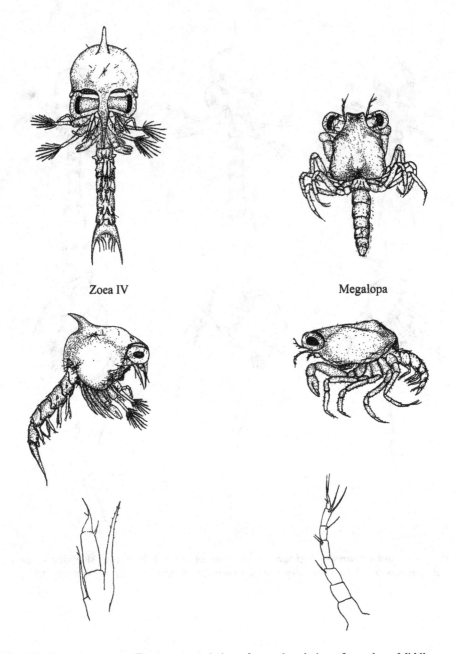

Zoea IV   Megalopa

Fig. 45b. *Sesarma cinereum*. Top row, ventral view of zoea, dorsal view of megalopa. Middle row, side view of zoea and megalopa. Bottom row, zoeal and megalopal antennae. [Reprinted with permission from Costlow & Bookhout, 1960.]

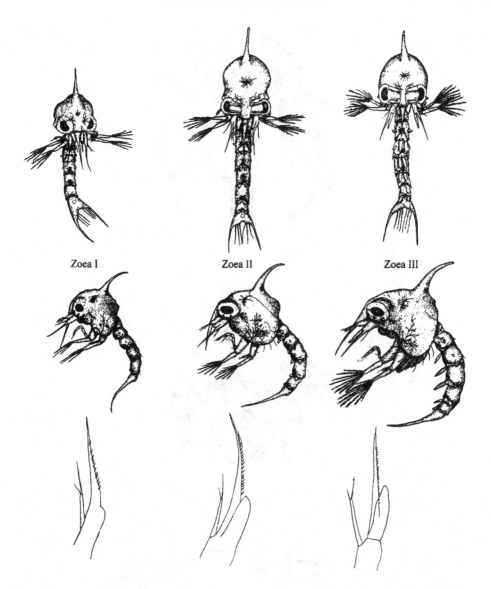

Zoea I   Zoea II   Zoea III

Fig. 46a. *Sesarma reticulatum*. Top row, ventral view of zoeae. Middle row, side view of zoeae. Bottom row, zoeal antennae. [Reprinted with permission from Costlow & Bookhout, 1962b.]

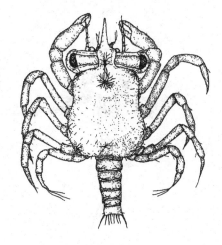

Megalopa

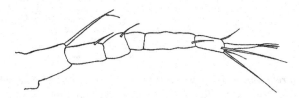

Fig. 46b. *Sesarma reticulatum*. Top row, dorsal view of megalopa. Middle row, side view of megalopa. Bottom row, megalopal antenna. [Reprinted with permission from Costlow & Bookhout, 1962b.]

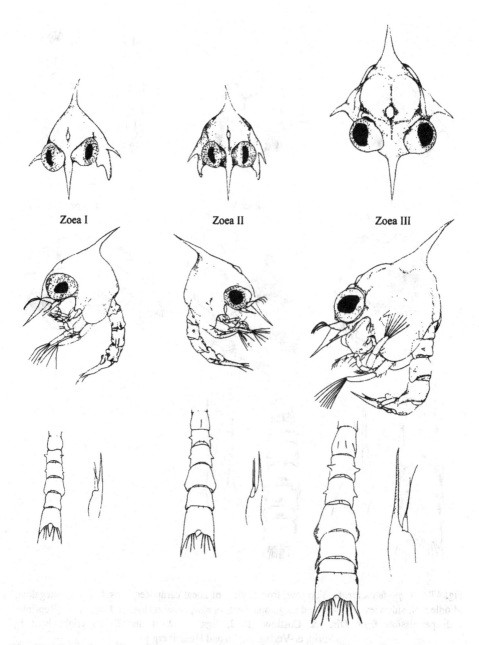

Zoea I   Zoea II   Zoea III

Fig. 47a. *Ocypode quadrata*. Top row, frontal view of zoeal carapaces. Middle row, side view of zoeae. Bottom row, zoeal telsons and antennae. [Reprinted with permission from Diaz & Costlow, 1972, figs. 1, 2 and 4; copyright held by Springer-Verlag, Berlin and Heidelberg.]

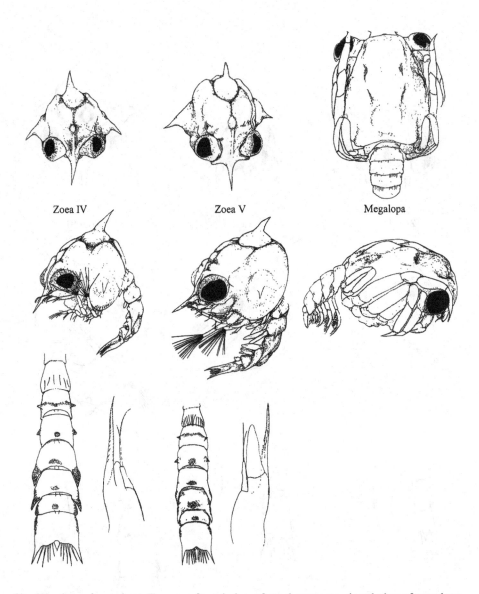

Fig. 47b. *Ocypode quadrata*. Top row, frontal view of zoeal carapaces, dorsal view of megalopa. Middle row, side view of zoeae and megalopa. Bottom row, zoeal telsons and antennae. [Reprinted with permission from Diaz & Costlow, 1972, figs. 1, 2, 4 and 9; copyright held by Springer-Verlag, Berlin and Heidelberg.]

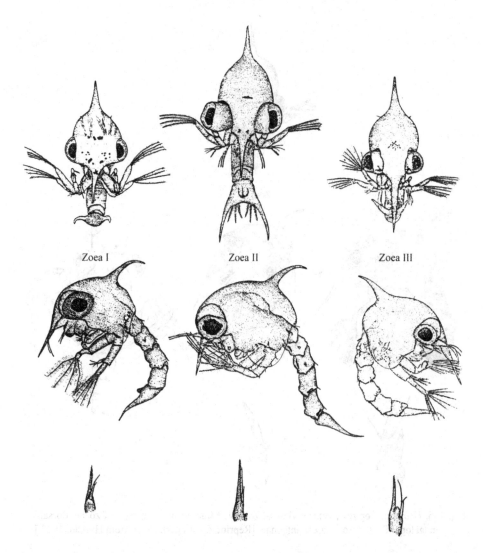

Zoea I   Zoea II   Zoea III

Fig. 48a. *Uca* spp. Top row, ventral view of zoeae. Middle row, side view of zoeae. Bottom row, zoeal antennae. [Reprinted with permission from Hyman, 1920.]

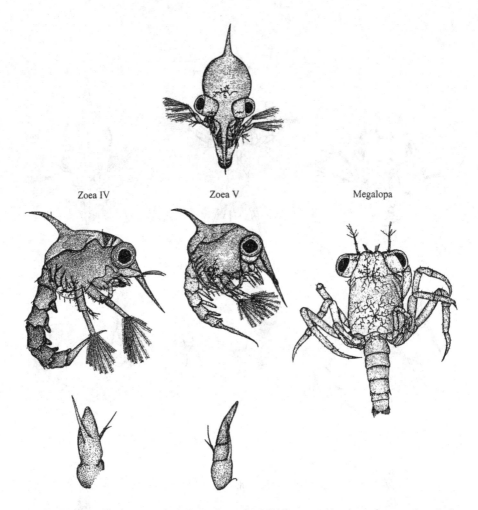

Fig. 48b. *Uca* spp. Top row, ventral view of zoea V. Middle row, side view of zoeae, dorsal view of megalopa. Bottom row, zoeal antennae. [Reprinted with permission from Hyman, 1920.]

TAXONOMIC INDEX